D1395136

Sensualities/Textualities and Technologies

Also by Susan Broadhurst and Josephine Machon

PERFORMANCE AND TECHNOLOGY: Practices of Virtual Embodiment and Interactivity

Also by Susan Broadhurst

DIGITAL PRACTICES: Aesthetic and Neuroesthetic Approaches to Performance and Technology

Also by Josephine Machon

(SYN)AESTHETICS: *Re*defining Visceral Performance

Sensualities/Textualities and Technologies

Writings of the Body in 21st Century Performance

Edited by

Susan Broadhurst and Josephine Machon

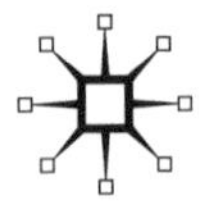

First published 2009 by
PALGRAVE MACMILLAN

Palgrave Macmillan in the UK is an imprint of Macmillan Publishers Limited, registered in England, company number 785998, of Houndmills, Basingstoke, Hampshire RG21 6XS.

Palgrave Macmillan in the US is a division of St Martin's Press LLC, 175 Fifth Avenue, New York, NY 10010.

Palgrave Macmillan is the global academic imprint of the above companies and has companies and representatives throughout the world.

Palgrave® and Macmillan® are registered trademarks in the United States, the United Kingdom, Europe and other countries

ISBN-13: 978-0-230-22025-6 hardback
ISBN-10: 0-230-22025-8 paperback

This book is printed on paper suitable for recycling and made from fully managed and sustained forest sources. Logging, pulping and manufacturing processes are expected to conform to the environmental regulations of the country of origin.

A catalogue record for this book is available from the British Library.

A catalogue record for this book is available from the Library of Congress.

10 9 8 7 6 5 4 3 2 1
18 17 16 15 14 13 12 11 10 09

Printed and bound in Great Britain by
CPI Antony Rowe, Chippenham and Eastbourne

Contents

List of Illustrations

Foreword

Ric Allsopp

> A path is a prior interpretation of the best way to traverse a landscape, and to follow a route is to accept an interpretation, or to stalk your predecessors on it as scholars and trackers and pilgrims do. To walk the same way is to reiterate something deep; to move through the same space the same way is a means of becoming the same person, thinking the same thoughts. It's a form of spatial theatre ...
>
> (Rebecca Solnit, 2002: 68)

> There are three beautiful things: bending down while remaining upright, a cry while remaining silent and a dance without moving.
>
> (Rebbe Mendel Kalish of Warka, in Unterman, 2008: 175)

The three central topics of this collection of chapters – bodies, technologies and texts – have constituted the broad conditions of theatre and performance at least since the period of classical antiquity. The relationship between these topics – how they interact with each other, how they might be identified as individual elements in a taxonomy of performance, what cultural values are inscribed in them or ascribed to them, and the variety of affects a particular distribution or arrangement of these terms might produce – is of course as much a focus of the history of poetic, performance and dramaturgical practice as it is of the chapters that follow here.

The classical (dramatic) view of theatre as a *theatron*, a 'seeing place', a prosthesis of eye and ear that privileged the 'objective' senses of sight and hearing through which it structured the vision of the audience, has increasingly given way to the more recent possibilities, driven by technological innovation, of performance as a site of immersive (as opposed to quasi-objective) sensual/sensory experience, an experience which is always produced and constructed through an inseparable relationship between bodies, texts and technologies.

The possibility of aligning the 'subjective' senses of touch and taste with the objective senses – of effectively merging bodies, texts and technologies – brings the whole human and post-human sensorium into play in ways that would have been unachievable and unthinkable until recently. And here it is useful to understand 'sensorium' as both the reciprocal effect of media on our senses in McLuhan's terms, and as our capacity to perceive and interact with the environments in which we live.

If from our contemporary perspective the classical (dramatic) view of theatre represents a predominantly fixed and normative set of relationships between bodies, texts and technologies that have remained largely operative until the present and certainly until the mid-twentieth century, then it is because in many senses they have been imagined and practised as discrete systems which allow or acknowledge a defined and disciplined set of interactions.[1] But each of these discrete systems (as McLuhan realized) also informs the condition and understanding of others in what is called a 'transductive' relationship.[2] We cannot, for example, disaggregate the effect of print technologies from our consciousness, as Walter J. Ong has pointed out, or perhaps the effect of controlled light from our imagination of performance, or effect of replay on our understanding of narrative. The acoustic technologies of classical theatre – the manipulation of architectural space to enhance the voice – were an integral part of a conception of a *polis* or what Ranciere refers to as a 'common term of measurement' (Ranciere, 2007: 44–5).[3]

The classical conception of the 'theatron' is also one through which consensual and idealized notions of *polis*, citizenship and community were produced. Whilst the conceptions of *polis* may have shifted, the idea(l) of a common place that might be produced through the interaction of bodies, texts and technologies remains. In the poet and theorist Lyn Hejinian's formulation, and with regard to a poetics of language, '... this common place is political in character. It constitutes *polis*. The political in this sense is not adjudicating and legislating but coming into appearance, and the *polis* is the space of appearance; the place for sharing words and deeds' (Hejinian, 2000: 366). The common measure shifts, and is shifted through, the mutable and fluid relationship of bodies, texts and technologies in any given context.

In her 1977 essay 'Modern theatre does not take (a) place', Julia Kristeva argued that theatre's 'only inhabitable place – locus – is language (*le langage*).' and noted that:

> [t]he Golden period of the Greek (or classical) community failed to materialize in the 20th century within existing theatrical communities, among totalitarians, fascist happenings and sociorealist pro-

> duction. As its only remaining locus of interplay is the space of language, modern theatre no longer exists outside of the text.
> (1997: 277)

This 'failure' of theatre to 'constitute a communal discourse of play (interplay)' as seen from a particular historical moment is perhaps countered by Hejinian's (1998) invocation of a 'space of appearance', which offers a less pessimistic view of the possibility of a 'communal discourse of play' with regard to the possibilities of both performance and language as sites of transformation.

That we no longer 'take (a) place' in the sense of locating a fixed or static relationship of bodies, texts and technologies, and that both the 'inhabitable space' and the 'space of appearance' that Kristeva and Hejinian identify, propose the possibility of a potential site of exchange, a moving alignment of text, sensation and technics, is evidence that in the context of an expanded field of performance, we are able to produce a generative and fluid space of inquiry, a 'thinking on' as Hejinian put it, that enables us to 'experience experience' and identify common measures. Language (by which I understand the interactivities of bodies, texts and technologies) is the medium that enables us to 'restore the *experience* of experience' and with it a 'sense of living our life' (Hejinian, 2000: 345). In the field of performance, bodies become other bodies, texts become other texts, and technologies become other technologies enabling the transformative and liquid flow of content across media. Bodies, texts and technologies then are no longer isolated as elements but are distributed, diffused and disseminated through performance. The body literally writes itself in performance, as Part 2 of this volume suggests. This sense of ephemeral inscription, of transformation, is no longer limited to bodies or texts transforming within a fixed or static scene or place. Recent technology enables a more complex interactivity between the elements that constitute the work, specifically through the digitalization of visual, graphic, sonic and textual data, which in turn effects our conception of bodies, texts and technologies.

I want to propose that a key relationship here is one of 'insideness', a relationship to bodies, texts and technologies which extends our engagement with and through performance beyond a static, distanced or objective set of relationships. Writing on poetry as an event in language, Gerald L. Bruns suggests that Hejinian's statement that 'the poem is open to the world' (2000: 43) means that 'language is not inside the poet as an innate grammar for generating sentences; rather the poet is inside language – inside, moreover, not as in a control centre but as in an environment ...'; and that 'the poet in this event does not so much use language as interact with uses of it' (Bruns, 2005: 30). As the language poets and others have argued and made evident in their work,[4] we are 'inside' language and to be inside language proposes that any text is 'a medium for experiencing experience', and that language is never separated from its contexts, but is at any given moment a 'constructedness', a play of, with and through language that involves both technics and bodies. As Heidegger noted '[w]e cannot ... view the life that we are in from the outside; we are always in the midst of it, surrounded by its details' (quoted in Hejinian, 2000: 363). We are continually inside language, inside the sensorium and inside technology, and as such constantly mediating and re-mediating our active environments and contexts.

The sense of being a generative and transformative part of the condition that we are in (a field of reference, a context, what gestalt psychology called a 'geography of requiredness') has formed a central condition of thinking about poetic form and 'open work' since at least the mid-twentieth century. But 'insideness' is also a condition, a modality of being, that is linked with being inside technology and inside the constructedness of bodies and texts in the sense of being in a constant transductive interaction with the world. We are not simply outside, in the sense often assumed in thinking about art or performance as forming the production of an object that lets us see into the condition of things. We are the condition of things and the 'real-time', and other interactive and digital technologies that we use (and which reciprocally use us) are in many senses what we become. This sense of 'insideness' that I am proposing here, of being a dynamic force in a given environment (what Hejininan would call being 'launched in to context') is analogous to the sense of the immersive which has always been a part or an aspiration of art experience – that we are taken up or taken beyond in the experience of the work. Its modality (and possibility) has begun to shift radically through more contemporary alignments of bodies, texts and technologies.

How can this 'insideness' be described? If in the situation of performance, we are inside language in Bruns's formulation of 'interacting with uses of it' rather than merely using it, then the condition of 'insideness' can be understood as a form of textuality. If a text is anything that can be read, from a gesture to a drawing, a dance to a poem, then textuality is *how* it is read and is therefore also a practice. To practise (as a way of operating or doing things in deCerteau's sense) is to be inside language and the varied modes of textuality (of how we interact with language) reflect the ways in which sensory modalities are stimulated: the affect of a particular poetics, or of a specific medium.

In a similar way, being inside bodies – a sensorium which incorporates the effects of media on the senses, and the sum of perceptions, which are always unstable, and constantly in a process of being activated or repressed – is also a practice, a state of textuality, of proprioception which the poet Charles Olson described as 'the 'body' itself as, by movement of its own tissues, giving the data of, depth' (Olson, 1974: 17). The question of how we are inside the body constantly returns us to the porosity of the body, its continual absorption and expulsion of what it is not, and thus to the sense of the environments and contexts which our bodies interact with as inscription and description. The body literally writes itself in performance: the dancer for example 'from one second to the next ... forms another body from his body ... [and b]y so doing he destroys what he was in order to attain what he is going to be' (Roger Lannes, 1938, quoted in Cramer, 2007: 11).

Likewise being inside technology is a condition of being human, which is to say that technologies are not the result of being human. As the philosopher Bernard Stiegler proposes 'it is the tool ... that invents the human, not the human that invents the technical' (quoted in Gere, 2006: 19). As with our being inside bodies, our understanding of the world depends on the technical means by which we apprehend it. 'It is through technics that the human is given access to the 'already there', to a past that he or she did not inhabit, and does not otherwise have access to ...' (ibid.: 19).

In as much then as we operate inside language, inside our bodies and inside technology, interacting with our uses of them, this 'insideness' (the sum of experiences that produce the subject) also means that we share a responsibility to understand the way in which we may be increasingly incorporated into triangulations of bodies, texts and technologies where the frames of reference move beyond or outside our control in productive ways or where objective freedoms become eroded. A responsibility to enable us to understand *how* these intersections and interactivities of bodies, texts and technologies are operating in the field of performance is the aim of this book.

The two quotations that serve as an epigraph suggest a poetic link between iteration and the ineffable, between placement and displacement, between bodies, texts and technologies seen as possibilities of language. The continual rethinking of this relationship, which is both the utopian and dystopian project of art and performance, is undertaken through putting the relationships of body, text and technology into question, through destabilizing and unsettling their tendencies to inertia, both as practice itself and through a critique of that practice, by inhabiting and activating the 'space of appearance'.

Notes

1 See Hans-Thies Lehmann (2006).
2 See Charlie Gere, 'Interiority and exteriority constitute the terms of what the philosopher Gilbert Simondon called a 'transductive' relation: a relation that constitutes these terms, meaning that a term in the relation cannot exist outside of that relation, and is constituted by the other term of the relation' (2006: 19).
3 Ranciere further argues that this 'common term of measurement' has subsequently been lost or replaced by the 'common factor of dis-measure or chaos that now gives art its power'.
4 See, for example, the work of Charles Bernstein, Allen Fisher, Lyn Hejinian, Rosmarie Waldrop, Barrett Watten.

References

Bruns, Gerald L. (2005). *The Material of Poetry: Sketches for a Philosophical Poetics.* Athens and London: University of Georgia Press.
Cramer, Franz-Anton. (2007). 'Knowledge, Archive, Dance', in *Capturing Intention.* Ed. Scott deLahunta. Amsterdam: Emio Greco | PC & AHK, pp. 11–14.
Gere, Charlie. (2006). *Art, Time and Technology*. Oxford: Berg.
Hejinian, Lyn. (1998). 'A Common Sense', in *The Language of Inquiry*. Berkeley: University of California Press, 2000, pp. 355–82.
——. (2000). *The Language of Inquiry*. Berkeley: University of California Press.
Kristeva, Julia. (1997). 'Modern Theatre does not take (a) place', in *Mimesis, Masochism & Mime*. Ed. Timothy Murray. University of Michigan Press.
Lehmann, Hans-Thies. (2006). *Post-Dramatic Theatre*. London: Routledge.
McLuhan, Marshall. (2001). *Understanding Media*. London: Routledge.
Olson, Charles. (1974). 'Priopriception', in *Additional Prose*. Ed. George F. Butterick. Bolinas: Four Seasons Foundation.
Ong, Walter J. (2002). *Orality & Literacy*. London: Routledge.
Ranciere, Jacques. (2007). *The Future of the Image*. London: Verso.
Solnit, Rebecca. (2002). *Wanderlust: A History of Walking*. London: Verso.
Unterman, Alan. (2008). *The Kabbalistic Tradition*. Harmondsworth: Penguin.

Notes on the Editors

Susan Broadhurst is a writer and performance practitioner and Professor of Performance and Technology in the School of Arts, Brunel University, West London. She is co-editor with Josephine Machon of *Performance and Technology: Practices of Virtual Embodiment and Interactivity* (Palgrave MacMillan, 2006), She is also sole author of *Liminal Acts: A Critical Overview of Contemporary Performance and Theory* (London: Cassell; New York: Continuum, 1999) and *Digital Practices: Aesthetic and Neuroesthetic Approaches to Performance and Technology* (Palgrave MacMillan, 2007), together with various articles in this area. She is co-editor of the *Body, Space & Technology* on-line journal <http://www.brunel.ac.uk/bst/>. Sue is currently working on a series of collaborative practice-based research projects, which involve introducing various interactive digital technologies into live performance including, artificial intelligence, 3D film, modelling and animation, and motion tracking and biotechnology.

Josephine Machon is a writer and practitioner in the creative arts and a Lecturer in Drama at Brunel University, West London. Josephine has recently authored *(Syn)aesthetics – Redefining Visceral Performance* (Palgrave Macmillan, 2009) and co-edited *Performance and Technology: Practices of Virtual Embodiment and Interactivity* (Palgrave Macmillan, 2006) with Susan Broadhurst. Josephine has been the Sub-Editor of *Body, Space & Technology* since its inception in July 2001. Josephine's past and present interdisciplinary research collaborations explore the fusion of embodied practice and the written text within *play*ful performance encounters.

Notes on the Contributors

Mojisola Adebayo is an actor, writer, teacher, director and producer. She also specializes in Theatre of the Oppressed. Mojisola has worked extensively with diverse international companies from The Royal Shakespeare Company, to Vidya, a slum dweller's theatre company in Ahmedabad, India. Her independent productions include *Moj of the Antarctic: An African Odyssey* and *Muhammad Ali and Me. Theatre for Development: A Handbook* co-written with John Martin and Manisha Mehta is published in 2009. Mojisola is writing *Matt Henson, North Star,* researched on an expedition to the Arctic with Cape Farewell and is also commissioned by Queer Up North and Nitro.

Ric Allsopp is a co-founder and joint editor of *Performance Research,* a quarterly international journal of contemporary performance (pub. London & New York: Routledge, Taylor & Francis) for which he has recently co-edited 'On Choreography' (2008) and 'Transplantations' (2009). He has been involved in the innovative Performance Writing programme at Dartington College of Arts since its inception in 1994 and was Director of Writing from 2001-2004. He is a Senior Research Fellow in the Department of Contemporary Arts at Manchester Metropolitan University, and currently a Visiting Professor at the Centre for Dance Education (HZT), University of the Arts (UdK), Berlin.

Phil Ellis is a digital artist and lecturer in Media Arts and Television Arts at the University of Plymouth. He recently completed an MPhil at the University of Plymouth entitled: 'An Investigation into the fluidity and stability of the "ScreenPage" in new media with particular reference to OuLiPo-ian techniques.' Recent works (that relate to the MPhil thesis and this chapter) *textimage generator* and *dada.doc* can be used at <http://www.philellis.plus.com/>. Phil is currently engaged in PhD research into affect and agency in relation to new television.

Rachel Fensham is Professor of Dance and Theatre Studies at the University of Surrey and Honorary Research Fellow at Monash University, Australia. Her research interests lie in the terrain between cultural aesthetics and politics, particularly in relation to creative processes and their reception. Her publications include works on performance theory, feminist and postcolonial theatre, as well as studies of cultural history and policy. Current research projects include mapping transnational and cross-cultural choreographies in Australia and archival research on women as modern dance pioneers in the UK.

Russell Frampton is a painter and film-maker, he is an Associate Lecturer in Theatre and Performance, University of Plymouth, teaching digital performance practice. Russell has an extensive background in researching the educational implications of the relationship between technology and performance through AHRC funded projects. He is an established visual artist exhibiting his paintings internationally and has devised both film and digital scenography for Lusty Juventus Physical Theatre company. In 2004 he formed 'Enclave Productions' with choreographer and dancer Ruth Way to extend his work into the arena of multimedia, film and collaborative practice.

John Freeman is a member of the Centre for Contemporary and Digital Performance at Brunel University, West London, where he is Senior Lecturer in Modern Drama Studies. In addition to numerous articles on contemporary performance, Freeman is the author of *Tracing the Footprints* (2003), *New Performance/New Writing* (2007) and *Blood, Sweat & Theory* (2009). His recent performance and research work, exploring relationships between identity-construction and issues of dislocation, migration and displacement form the basis of a subsequent book, *Forever/Elsewhere* for publication in 2012.

Sara Giddens (co-director) is a choreographer. She works as a teacher, lecturer, facilitator, researcher, mentor and project manager. Using interdisciplinary approaches and working independently and collaboratively with other artists and educators, particularly dancers, writers, sonic artists, architects, video/new media-makers, teachers, heads and creative organizations, her extensive national and international work seeks to challenge both the boundaries of conventional dance and performance and how those works can be documented and disseminated. In 2005, Sara was awarded a research bursary from the Choreographic Lab at the University of Northampton. She has published in *Liveartmagazine, Performance Research,* and *Feminist Review,* and most recently for *Creative Partnerships* through Engage: This is My Gallery, and was visiting artist at Banff Arts Centre (Canada, 2008).

Simon Jones (co-director) is a writer and scholar. He has written plays and texts for a number of companies since the 1980s. He has taught theatre studies at Lancaster University, and is currently Professor of Performance at the University of Bristol. He has been a visiting scholar at Amsterdam University (2001), a visiting artist at The School of the Art Institute of Chicago (2002) and Banff Arts Centre (2008), and worked with Spell#7 Performance (Singapore) in a revival of his performance text *Beautiful Losers* (1994/2003). He has published in *Contemporary Theatre Review, Entropy Magazine, Liveartmagazine, Performance Research, Shattered Anatomies* (ed. Heathfield and Quick), and most recently in *The Cambridge History of British Theatre*.

Roberta Mock is Reader in Performance at the University of Plymouth where she also leads the MRes Theatre & Performance. Her books include *Jewish Women on Stage, Film & Television* (Palgrave Macmillan) and as editor, *Walking, Writing & Autobiography* (Intellect), *Performing Processes: Creating Live Performances* (Intellect), and *Performance, Embodiment & Cultural Memory* (with Colin Counsell, Cambridge Scholars Publishing).

Stelarc explores alternate, intimate and involuntary interfaces with the body. He has performed with a Third Hand, a Virtual Body and Exoskeleton, a six-legged walking robot. He is surgically constructing an ear on his arm that will be internet enabled, making it publicly accessible to people in other places. He was appointed Honorary Professor of Art and Robotics at Carnegie Mellon University. He also was awarded an Honorary Doctorate by Monash University. He is currently Chair in Performance Art at Brunel University, West London, and is Senior Research Fellow in the MARCS Labs at the University of Western Sydney.

Dawn Stoppiello is a choreographer who has dedicated her career to computer mediated live performance. She received her BFA in dance from California Institute of the Arts and was a member of the Bella Lewitzky Dance Company. With composer Mark Coniglio she is co-founder of Troika Ranch with whom she has performed, lectured and taught in Australia, Western Europe, Canada and throughout the United States. Stoppiello received a 2004 Statue Award from the Princess Grace Foundation for her sustained achievement in dance. Company honors include a 2003 'Bessie' Award and an Honorable Mention from the 2004 Prix Ars Electronica Cyberarts Competition.

Olu Taiwo is Senior Lecturer in the Faculty of Art at the University of Winchester. His post-doctorial studies are concerned with performative nature regarding the relationship between 'effort', 'performance' and the 'performative' in different arenas. He aims to propagate twenty-first-century issues concerning the interaction between body, identity, audience and technology. Publications include *The Return Beat,* in *The Virtual Embodied,* ed. Wood, (Routledge, 1998); *Music, Art and Movement among the Yoruba,* in *Indigenous Religions* (Cassell, 2000); *The Orishas: The Influence of the Yoruba Cultural Diaspora',* in *Indigenous Diasporas and Dislocations,* ed. Harvey and Thompson (Ashgate, 2005).

Fiona Templeton is a writer, performer and director of the New York based performance group The Relationship. Her award-winning productions recently include the epic *Medead* on Governor's Island; *L'Ile,* staging the dreams of the people of Lille, France, in the places dreamed of; *GOING (with Coming),* a recreated 1970s piece by the Theatre of Mistakes, of which she was a core member. Books include *YOU-The City* (an intimate Manhattanwide play for an audience of one), *Cells of Release* (Roof Books), *Delirium of Interpretations* (Green Integer), *Elements of Performance Art.* She teaches at Brunel University, London. <www.fionatempleton.org>; <www.therelationship.org>

Tracey Warr is a writer based in Wales. She is Lecturer in Contemporary Art Theory at Oxford Brookes University and also contributes to teaching at Bauhaus University, Weimar and University College Falmouth. Her recent publications include *Setting the Fell on Fire: Allenheads Contemporary Arts* (Editions North, 2009) and essays in *Half Life* (NVA, 2007); *Panic Attack!: Art in the Punk Years* (Merrell, 2007); *Marcus Coates: The Dawn Chorus* (Picture This, 2007); *Little Earth* (London Fieldworks, 2005) and *Joan Jonas* (John Hansard Gallery, 2004).

Ruth Way is Senior Lecturer and Head of Theatre and Performance, University of Plymouth, and Programme Leader for BA Dance Theatre. Ruth has had an extensive career as a dancer and choreographer performing with Earthfall and Lusty Juventus Physical Theatre and has presented her practice as research through performative papers, articles and performances internationally. With Russell Frampton she is co-director of Enclave Productions and their most recent film, *Utah Sunshine,* has been selected for screenings in France, Poland and Brazil. Current research is exploring connections between somatic movement practice and the sentient body in her performance and film practice.

Paul Woodward is a Senior lecturer in Drama and Physical Theatre at St Mary's University College. He has worked as a director/performer/writer for a variety of physical/experimental theatre companies in the early 1990s. Graduating with an MA (distinction) in Theatre at Royal Holloway, he consolidated his research into body/sign systems in Theatres of Asia and its application to Sign Language Theatres of the Deaf. Paul remains active as a professional director/dramaturg and has collaborated with Maxine Doyle (First Person dance/theatre) and Dr Josephine Machon (Brunel University) investigating the interface between the body, popular cultures and technology. He has delivered physical theatre workshops nationally and internationally, including the international festival of therapy and theatre, Lodz, Poland, and in Knysner, South Africa, working with HIV positive children in the townships. Paul is currently working on a practice as research PhD investigating the performativity of HIV (dis)closure.

Introduction

Susan Broadhurst and Josephine Machon

Sensualities/Textualities & Technologies: Writings of the Body in 21st Century Performance explores the interrelation of text, body and technology in performance. Our intention is to provide an edited collection which reconsiders textual practices in contemporary performance specifically focusing on the exciting exchange between writerly texts, body and technology. This collection adds to our previous co-edited publication, *Performance and Technology: Practices of Virtual Embodiment and Interactivity*, which placed emphasis on the corporeal in arts practice in this technological age. It also adds a further dimension to Palgrave Macmillan's series on Performance and Technology, which includes *Digital Practices: Aesthetic and Neuroaesthetic Approaches to Performance and Technology*. Crucially, the chapters in this book include analysis of important textual practices that are central to contemporary performance. In particular, they offer an interrogation of diverse 'textualities', that is, performance writing that is indebted to sensual 'writings of the body' in all manner of ways. At the same time the collection looks to new approaches offered by ongoing textual practices in physical, visual and virtual performance that incorporate new and existing technologies.

This collection is divided into sections that provide an overview of a particular area in relation to writings, the body and technology. These are: Part 1 'Writing and Technologies – New Epistemologies'; Part 2 'The Body Writes Itself ...'; Part 3 'Performing the Body/Performing the Text ...Writing the Body/Writing the Text'; and Part 4 'Corporeal Intertextualities – Body/Text/Technologies'. To discuss these areas from a rich variety of perspectives we have assembled an international collection of writers, performers and academics renowned in their field, each of whom prioritize sensualities, textualities and technologies within their own thinking and practice.

In exploring textualities a number of our contributors have chosen to play (*jouer*) with format and writing style in order to draw the reader into an active engagement with the text on the page as much as the ideas under scrutiny, in this way emphasizing their writerly approach at the heart of this book. For instance, Sara Giddens and Simon Jones and also Stelarc play with layout including various sizes and styles of fonts; whereas Mojisola Adebayo and Fiona Templeton explore the performative nature of writing which seeks to (re)present the live performance process central to their own work. This writerly play adds texture and diversity to the reader's sensual experience of *Sensualities/Textualities & Technologies*.

Susan Broadhurst opens the debate regarding 'Writing and Technologies – New Epistemologies', showing how, in a relatively short period of time, there has been an explosion of new technologies that have infiltrated all areas of life and irrevocably altered our lives. Broadhurst draws attention to the complexities thrown up when the meaning aimed at cannot be reached by the body alone, it builds its own instruments and projects around itself a mediated world. Rather than being separate from the body, technology becomes part of that body and alters and recreates our experience in the world. Moreover, in many such performances the human body is shown in flux, a body where contacts are made not physically but electronically, where cutting-edge choreography and multimedia effects explore the inherent tensions between the physical and virtual; the body literally writing itself in performance.

Tracey Warr's 'Texts from the Body' (Chapter 2) examines Bruce Gilchrist's work, which uses a range of technologies to draw 'texts' from the mute body and to explore the unconscious, unlanguaged parts of consciousness. Gilchrist's work is discussed alongside William James's writings on the 'stream of consciousness' in the nineteenth century. With screen avatars, smart border biometrics, cosmetic surgery, the human genome project, genetic modification, and a borderless and id-less cyberspace, the twentry-first-century body is enmeshed in contradictions of fixity and flux, authenticity and simulation. Exactly what Gilchrist's body texts are trying to say to us is on the tips of our tongues.

Warr's chapter is followed by Sara Giddens and Simon Jones of Bodies in Flight. Chapter 3 examines how this British performance company uses a combination of technologies to explore the status of human identity in contemporary culture. Via ongoing intermedial collaborations, Bodies in Flight progressively interrogate the impact of digital technologies on the sense of self and human interrelationships. They highlight how these technologies can be used in performance to expose the intimate process of incorporation into the human psyche, or, what Bodies in Flight call 'second-naturing'.

'Writing and Technologies – New Epistemologies' is drawn to a close with Chapter 4 by Phil Ellis. This chapter explores the effects of the shift from the linear printed text into its integration within digital media, focusing on the creative possibilities of user interaction inside the 'ScreenPage'; a term used by the author to identify the relationship between our understanding of the convergence of the linear page with multiple types of interactive media and their emergent language form. The centrality of the reader/user and the mediating technology in the relationship between author and work in interactive new media becomes pivotal.

Our next section is Part 2 'The Body Writes Itself', which is opened by international performance artist, Stelarc, who presents his recent project, 'PROSTHETIC HEAD' in Chapter 5. The Prosthetic Head is an embodied conversational agent (ECA) that speaks to the person who interrogates it. The interface is a keyboard and the text box beneath the projected head confirms the user's query. Exactly how questions are asked will determine the Head's response. Therefore, there is a translation between the keyed in text and the text-to-speech engine that coupled with the geometry and animation of the 3D model results in real-time lip synching and spoken response. The effectiveness and seductiveness of the ECA depends on it being convincing in both its comprehension and communication with the user. The Head, with its facial behaviour and basic repertoire of expressed emotion performs with words. The Prosthetic Head then is a conversational system that when coupled to a human user is capable of some interesting, often appropriate and at times unpredictable exchanges. Moreover, the Head at times resorts to speaking both bits of information from its data-base and speaking bits of its code thus exposing its own programming.

Paul Woodward's 'Performative (Dis)closures – Sensual Readings and Writings of the Positive Body' explores selected examples of live performance as a way of interrogating the tensions prevalent between the medical body and the autobiographical body within live performance. Through the analysis of a range of performance artists active in this field, including Woodward's own ongoing performance project *The Healing Room*, Chapter 6 investigates the ways in which the bodies of HIV+ men, once diagnosed, undergo a journey of personal and public surveillance. By examining the significance of live performance in reconciling the outer/inner world of the HIV+ performer, Woodward considers the role the audience might play in this journey and surveys the interplay between written text and performative action within this scenario, highlighting the interrelatedness of aspects of performativity in the everyday life of the positive body and the performance of sero-status in a theatrical context.

In the consideration of the many ways in which the body writes itself, Mojisola Adebayo's 'The Supernatural Embodied Text: Creating *Moj of the Antarctic* with the Living and the Dead' examines how biography can become fiction and then become auto/biography; how historical text can become fantastical and at once again historical, within the creation of a physicalized performance text. It highlights through each of these moments how the corporeal experience of writing takes on qualities of the supernatural. Employing key moments in this multimedia performance as illustration, Chapter 7 explores the embodied exchange that occurs in the diverse acts of writing throughout the performance process.

Part 2 closes with Chapter 8, Olu Taiwo's 'Physical Journal: The Living Body that Writes and Rewrites Itself'. Here Taiwo seeks to advance a way of conceptualizing living bodies from the perspective of a practitioner. According to Taiwo, the proposed concepts, underpinned by embodied memories, are from lived experiences informed by movements derived from his body in its process of change. These memories neurologically construct a kind of virtual body, which is holographically projected as spatialized memory helping a living body to write and rewrite itself. He highlights how, from the movements of an embryo, to the motion of the last breath, there is left a movement journal, a lived duration, a series of events imprinted in temporal spaces recorded in the time-lines of human ephemeral and subjective universes. Physical journals exist as sets of holographic memories that aid in manifesting lived bodies.

Our collection then moves in Part 3 into a consideration of ideas and practices around 'Performing the Body/Performing the Text ... Writing the Body/Writing the Text', beginning with John Freeman's 'Socialising the Self: Autoethnographical Performance and the Social Signature'. Freeman argues in Chapter 9 that 'postmodern performance', a style that he believes is already over, was both built on narrative forms, as well as subverting these traditional forms in its drive for resistance. Freeman suggests that the boundary between these forms and other types of theatrical performance became diffused, creating a liminal space, or void, into which increasingly formulaic performance was cast. As he posits within this chapter, it is within this space that autoethnographical performance is developing as a form of 'post-postmodernism' that is 'postmodernism with a social conscience'.

Roberta Mock in Chapter 10 discusses concepts surrounding 'Deviant Textualities and The Formless' as illustrated by La Fura dels Baus's *XXX*. She shows how the Marquis de Sade's novel *Philosophy in the Bedroom* was reworked in *XXX* by employing a range of technological devices in order to construct a performance text written through and as somatic possibility. Mock suggest this presents a challenge to authenticity, as she interrogates through George Bataille's concept of the *informe* (or the formless).

Certain ideas put forward by Mock are pertinent to Rachel Fensham's chapter, which explores the notion of 'bodies in suspension' as she considers the aesthetics of doubt in the Australian Dance Theatre's *Honour Bound*. Fensham argues in Chapter 11 that performance can approach the representation of torture by making bodily suspension the corollary of doubt. Focusing particularly on the visceral choreography of Garry Stewart, she examines the way in which different kinds of texts, personal testimony, the Geneva Convention on Torture, and a virtual tecnnologized environment, suspend the performer's bodies in a state of liminality. Fensham suggests that the manipulation of legal and political texts has delimited recent discourses on torture.

Dawn Stopiello's Chapter 12 'Translation: Words < – > Movement < – > Bits' closes this section. She posits that her own choreography cannot be made without words. Informed by decades of journal writing, dancing and working with computer systems, Stoppiello's creative practice is a fluid interchange of language, movement and 'bits'; her choreographic process always begins with words and the alphabet has served as an algorithmic system of generating choreography from texts. Stoppiello's chapter is an exposé of how her hybridized practice with the intermedia company Troika Ranch translates words/language/text into movement.

The final section in this collection considers the manifold 'corporeal intertextualities' between Body/Text/Technologies. Part 4 begins with Fiona Templeton's 'Speaking for Performance/Writing with the Voice'. Here Templeton describes and discusses techniques that she has used within her own practice of 'creating text for performance without writing'. She suggests in Chapter 13 that speech that imitates writing is 'papery'. Following a sensual and performative writing style, Templeton muses on her own practice and describes how writing for performance is texted on the page to be recreated outside of that space, outside of space, in time, spoken. She shows how creating for speaking, through, from and for performance, but without writing, is texted upon, with and for time, by the body; the body always in time. Unlike the word on the page or the reading mind, speaking for performance cannot avoid being simultaneous with its co-textual speeches.

Ruth Way and Russell Frampton in Chapter 14 follow this discussion in a similarly performative vein. Their chapter focuses on layering authenticity and perception and the 'blurring spaces' between personal space, theatrical space, external spaces and selected archive footage in the making of their performance of *Utah Sunshine*. Their work seeks to avoid any sense of a linear narrative, instead the creative processes instil and craft a set of creative ambiguities that aim to invite other ways of seeing and experiencing these texts.

Part 4 and the collection concludes with Chapter 15, Josephine Machon's '(Syn)aesthetic Writings: Caryl Churchill's Sensual Textualities and the Rebirth of Text'. Here Machon surveys contemporary play-writing that is textured with interdisciplinary practice within the fabric of the text. Machon argues that the idiosyncratic fusions of body/text/technology, which are present in the very fibres of Churchill's texts, point towards a rebirth of writerly textualities in current theatre practice. Applying Machon's (syn)aesthetics as the defining style and strategy of appreciation of such work, Chapter 15 examines the experiential quality of these sensual textualities in this technological age.

To conclude, as each section heading will make the reader aware there are themes and concepts that run through each one. The reader may also enjoy further connections that intertwine within and across the sections, finding additional pathways through the collection. As regards theorization and conceptual connections, various ideas around intertextualites and embodied textual practice, such as Georges Bataille's 'informe' and Maurice Merleau-Ponty's delimited body, are echoed throughout a number of chapters. As highlighted above various authors explore a performative style within their analysis, further reflected in their style and format of presentation. This plays on the sensual practice of the body writing itself, which is foregrounded in those chapters that examine and expound notions of autobiography and autoethnography. Overall, we encourage the reader to delight in the sensual play that exists within this anthology between the body, text and technology.

Part 1

Writing and Technologies – New Epistemologies

1
Digital Practices: New Writings of the Body

Susan Broadhurst

To be a body is to be tied to a certain world.
(Merleau-Ponty, 1962: 148)

In digital practices, instrumentation is mutually implicated with the body in an epistemological sense. The body adapts and extends itself through external instruments. To have experience, to get used to an instrument, is to incorporate that instrument into the body. The experience of the corporal schema is not fixed or delimited but extendable to the various tools and technologies which may be embodied. Our bodies are always open to and intertwined with the world. Instruments appropriated by embodied experience become part of that altered body experience in the world. In this way, 'the body is our general medium for having a world' (ibid.: 146).

Technology, then, would imply a reconfiguration of our embodied experience. When the meaning aimed at cannot be reached by the body alone, it builds its own instruments and projects around itself a mediated world. Rather, than being separate from the body, technology becomes part of that body and alters and recreates our experience in the world. Moreover, the body is a system of possible actions since when we point to an object, we refer to that object not as an object represented, but as a specific thing towards which we 'project' ourselves (Merleau-Ponty, 1962: 138), in fact a 'virtual body' with its phenomenal 'place' defined by task and location (ibid.: 25). This emphasis on a virtual body has resonance with and points to a deconstruction of the physical/virtual body of digital practices, a body of potential creativity.

Magnetic or optical motion capture is exemplary of this 'instrumentation' and has been used widely in performance and art practices for some time now. This involves the application of sensors or markers to the performer or artist's body. The movement of the body is captured and the resulting skeleton has animation applied to it. This data projected image or avatar then becomes some part of a performance or art practice. Motion tracking is used especially in live performances, such as Cunningham's *Biped* (2000), where pre-recorded dancing avatars are rear-projected onto a translucent screen, giving the effect of a direct interface between the physical and virtual bodies; and in digital sound and new media interactive practices, where during Troika Ranch's *Surfacing* (2004), captured live or pre-recorded images freeze, fragment, speed up, slow down or warp in a shimmering effect – all by means of Isadora software (Figure 1.1).

In Stelarc's performances the body is coupled with a variety of instrumental and technological devices that, instead of being separate from the body, become part of that body, at the same time altering and recreating its experience in the world. One such performance is *Muscle Machine* (2003), where Stelarc constructed an interactive and operational system in the form of a walking robot (Figure 1.2). This intertwining of body, technology and world is important since instead of

Figure 1.1 Danielle Goldman in *Surfacing* (2004). Photo: Richard Termine

Figure 1.2 *MUSCLE MACHINE*, Gallery 291, London (2003). Photographer: Mark Bennett. STELARC

abandoning the physical body, instrumentation and technology extends it by altering and recreating its embodied experience.

'Performance and technology' in its various forms is an emergent area of performance which reflects a certain being in the world. In a relatively short period of time there has been an explosion of new technologies that have infiltrated all areas of life and irrevocably altered our lives. Consequences of this technological permeation are both ontological and epistemological, and not without problems as we see our world change from day to day.

My main premise concerns the exploration and investigation of the physical/virtual interface so prevalent within the digital. In my opinion, this area has not been adequately theorized as yet and as a result poses a number of questions. For instance, does this new physical/virtual interface give rise to a new aesthetics? If so, what then are the theoretical and practical implications of this? It is my intention to explore and analyse the effect these new technologies have on the physical body in performance. As a development of my previous theorization on liminality (Broadhurst, 1999a, 1999b, 2004b), I believe that aesthetic theorization is central to this analysis.[1] However, other approaches are also valid, particularly, those offered by recent research into cognitive neuroscience,[2] particularly in relation to the emergent field of 'neuro-aesthetics' where the primary objective is to provide 'an understanding of the biological basis of aesthetic experience' (Zeki, 1999: 2).

In the digital there is a proliferation of performances that utilize electronic sound technology for real-time interaction. One performance group who explore the use of this technology is Optik, who have performed at various national and international venues and now prioritize the use of digitally manipulated sound in their movement-based performance.

Sound technologies central to their performances are MIDI (musical instrument data interface) and Max, a real-time programming environment that has the special advantage of being interactive with visual and network technologies. Established as an agreed universal standard method for sending and receiving musical controller information digitally, the application of the basic MIDI interface has expanded in a variety of ways. It provides a standardized interface for a wide variety of control devices. Its codes have also been adapted to control various non-musical devices to coordinate with video and graphics. The development of MIDI has had a strong impact on the accessibility and variety of interactions that can be utilized in performance However, the accompanying necessary restriction of data means that important musical information is lost and the processing speed of MIDI is restrictive for real-time interaction. This in turn has led to the proliferation of numerous highly programmable interactive MIDI systems that can offer immediate feedback, Max being the most widely used. It has also led to the development of OSC (Open Sound Control),[3] which is a protocol that allows the real-time control of computer-synthesis processes from gestural devices.

MIDI, Max and OSC are central to the performances of Troika Ranch, who fuse traditional elements of music, dance and theatre with real-time interactive digital technology. They are pioneers in their use of MidiDancer and Isadora software, which can interpret the physical movements of performers, and as a result that information can be used to manipulate the accompanying sound, media and visual imagery in a variety of ways, thus providing a new creative potential for performance.

Likewise, Palindrome, who focus on the interface and interaction between virtual sound and the physical body, also utilize such sound technologies. Artistic directors Robert Weschler and Frieder Weiss have designed and developed interactive software and hardware, including Eyecon, a camera-based motion-sensing system. Their choreography is affected by the live generation of sound through the use of sensors and real-time synthesis, and those movements in turn shape the resulting music.

When looking at objects in performance, colour is perceived before form that in turn is perceived before motion (Zeki, 1999: 66). The consequence of this is that over very short periods of time, the brain is unable to combine what happens in real time; instead, it unifies the

Figure 1.3 *Solo4>Three* (2003). Dance and Choreography: Emily Fernandez. Interactive video system: Frieder Weiss

results of its own processing systems though not in real time. Nevertheless all visual attributes are combined to provide us with an integrated experience. Palindrome's shadows performances, as a result of their multi-layered, distorted and delayed effects, challenge this 'integrated experience' (Figure 1.3), at the same time they ensure the audience's active participation in the production of meaning. The shadows shift seamlessly between what is '*known*' and what is '*surprising*' making 'the piece fascinating to watch' (Dowling, Wechsler and Weiss, 2004: 5).

Troika Ranch's *The Future of Memory* (2003) explores memory and the act of remembering – 'how memories are created, stored, romanticized, repressed and lost' – by means of a multi-layered collage of imagery and sound; the technology acting as a 'metaphor for memory' itself (Coniglio and Stoppiello, 2005). Metaphor has been identified with the Freudian notion of 'condensation' and metonymy with 'displacement' (Lyotard, 1989). However, for Lyotard, there is a certain futility in bring everything back to the linguistic 'as the model for all semiology',

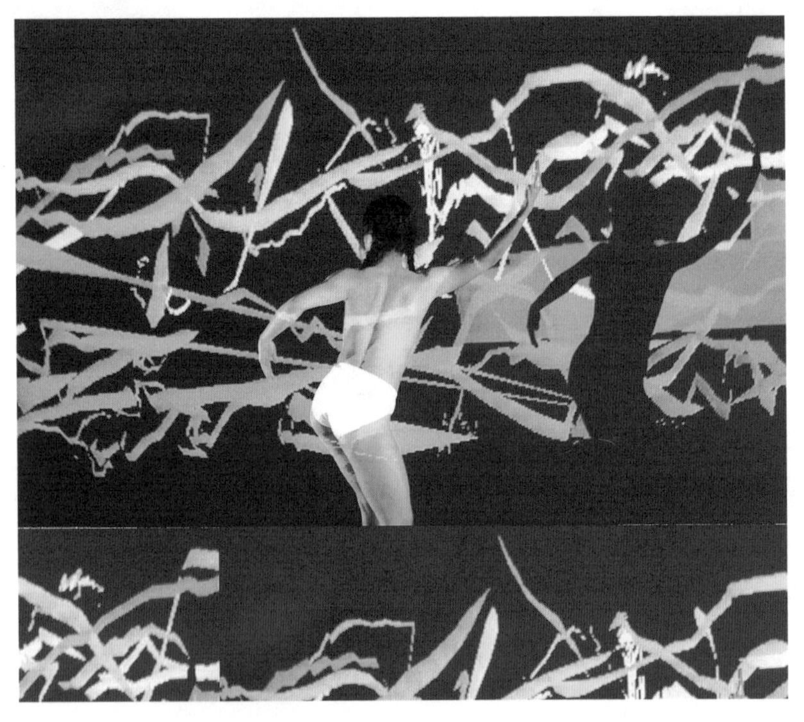

Figure 1.4 Traces of the performer's hands and feet leave multiple curved white traces, in *16 [R]evolutions* (2006). Performer: Lucia Tong. Photo: Richard Termine

when it is 'clear that language, at least in its poetic usage, is possessed ... by the figure' (ibid.: 30). The figural, is not the figuration of representational art but is instead that of creativity and elusiveness, and it is important since it mirrors many digital practices, placing the performance within the context of a libidinal economy.

A more recent work, *16 [R]evolutions* (2005) is a performance where cutting edge choreography and multimedia effects explore the similarities and differences between human and animal and the evolutions that both go through in a single lifetime; the body literally writing itself in performance (Figure 1.4).

In my opinion such quintessential features demand a new mode of analysis which foregrounds the inherent tensions between the physical and virtual. These practices, in different ways, emphasize the body and technology in performance. Therefore, my main premise is the exploration and investigation into the physical/virtual interface so prevalent within the digital.

Such works present innovation in art practices, being at the cutting edge of creative and technological experimentation. It is my belief that tensions exist within the spaces created by the interface of body and technology and these spaces are 'liminal' in as much as they are located on the 'threshold' of the physical and virtual. I am suggesting that it is within these tension filled spaces that opportunities arise for new experimental forms and practices. As such, I identify certain features that are central to these new practices. First and foremost, the utilization of the latest digital technology is absolutely central since within these various art practices and performances there is an assortment of technologies employed. Another important trait is an accentuation of the corporeal in terms of both performance and perception with its emphasis on intersemiotic modes of signification,[4] since in much of this performance the body is primary and yet transient.

Artificial intelligence is also featured in these technological practices, where the challenge is to demarcate the delimited human body from an artificially intelligent life form, such as Jeremiah the avatar from *Blue Bloodshot Flowers* (2001) who was developed from surveillance technology. One of the most interesting aspects of this performance is how much the performer/spectator projects onto the avatar. This is not so surprising since a substantial area of the human brain is devoted to face recognition (Zeman, 2002: 216).

The ability of humans to recognize facial expressions is so sophisticated that even very slight differences are noticed and made meaningful

Figure 1.5 Elodie Berland and Jeremiah from *Blue Bloodshot Flowers* (2001). Image by Terence Tiernan

and that is why faces such as Jeremiah's (Figure 1.5) have such a powerful effect on the spectator.

Similarly, Stelarc aims to introduce artificial intelligence into his *Movatar* prosthesis in order for it to perform in the real world, thus further blurring the distinctions between the virtual and actual. According to Stelarc, his avatar can be thought of as 'a kind of viral life form' or agent that lies dormant except when it is connected to a physical body, which causes it to become activated and it in turn reactivates the host body. Therefore, the body shares its agency with an artificial entity that has the ability to learn, developing its behaviour within the duration of a performance (2002: 76).

In digital practices, virtual bodies that are generated by physical movement through the mediation of digital technology are seen together with live performers. The performances with their interface and interaction between physical and virtual bodies can be seen to displace fixed categories of identity; each carries a 'trace' of the other, given that the virtual performers are the digital reincarnation of the human bodies. However, limits of the embodied self are not fixed since embodied emotional response can also be due to the stimulation of external objects that have been appropriated by the body (Ramachandran and Blakeslee, 1999: 61–2). Digital practices, with their use of Mocap (motion capture) and artificially intelligent technologies take this appropriation further since the motions of a performer's body captured technologically featuring avatars, such as Jeremiah and the movatar, results in a modified extension of that physical body. The implication being that the embodied self as any other aspect of the conscious self is transitory, indeterminate and hybridized.

Likewise, for Lyotard, instead of a conceptual interpretation of meaning, the 'figural', a territory of form, colour and the visual, indicates flows and drives of 'intensities' which continually displace the identity of the reader (or spectator) (Lyotard, 1988: 277). The aim is to eliminate oppositions in favour of intensities, and it is useful in a description of such digital practices as *Blue Bloodshot Flowers* with its diverse elements that often escape meaningful interpretation.

Other aesthetic features within the digital are heterogeneity, indeterminacy, fragmentation, hybridization and repetition. Due to the hybridization of these practices and the diversity of media employed, various intensities are at play. It is these imperceptible intensities, together with their ontological status that give rise to new modes of perception and consciousness. Central to becoming and making new connections is the body without organs or BwO. It is 'the field of immanence of desire' (Deleuze and Guattari, 1999: 191) and an 'intense and intensive body' (Deleuze, 2003: 44). Desiring machines and the body without organs can be seen as two sides of the same coin, or 'two states of the same 'thing', a functioning multiplicity one moment, a pure, unextended zero-intensity the next' (Bogue, 1989: 93). In Troika Ranch's performance of *The Electronic Disturbance* (1996), the ebb and flow between the organic and electronic is in a continual process of becoming and making new connections.

'Defamiliarizing' devices are also employed within digital practices, such as the juxtaposition of disparate elements that, in creating a distancing effect, causes the audience to actively participate in the activity of producing meaning. The employment of wide, jarring metaphors is another central characteristic of the digital. The colourful and figurative use of aural and visual imagery and the juxtaposition of metaphors evoke surreal dreamscapes. The interaction of the physical and virtual also creates inclusive, jarring metaphors. This mixture produces an aesthetic effect caused by the interplay of various mental sense-impressions, which unsettle the audience by frustrating their expectations of any simple interpretation and in turn produce a new type of synaesthetic effect that is analogous to the experience caused by cross-wiring or cross-activation of discrete areas of the brain in certain perceptual disorders (Ramachandran and Hubbard, 2001: 9).

Other forms of digital practices are those that incorporate biotechnology within their creative experimentation. Such art works are commonly referred to as 'Bioart'. The Tissue, Culture and Arts Project are such a group, whose tissue engineering exploration, exemplified by Stelarc's *Extra Ear* (2004), is integral to their art installations, resulting in works of varying geometrical complexity thereby creating a living 'artistic palette'.

Marta de Menezes is an artist who also works with biotechnology. For her project *Nature?* (1999), she reprogrammed patterns on butterfly wings by injecting the pupa in development. These pattern transformations relate only to the phenotype and not genotype and disappear at the end of the life cycle. She has also applied various colours to elementary parts of brain cells and through projections in 3D has created live sculptures. Her work *Functional Portraits* (2002) employs functional magnetic resonance (fMRI), which visualizes in real time the operation of the brain (Figure 1.6). In so doing de Menezes attempts to demonstrate the 'neuronal correlate of consciousness', which generally refers to the correlation between neuronal activity and the sensation, thought, or action that relates to that mental activity (Crick, 1994: 208).

The bioart of de Menezes is mirrored in Lyotard's notion of the figural as not the figuration of representational art, but instead of creativity and elusiveness, whilst at the same time placing this work within the context of a libidinal economy. A libidinal economy is

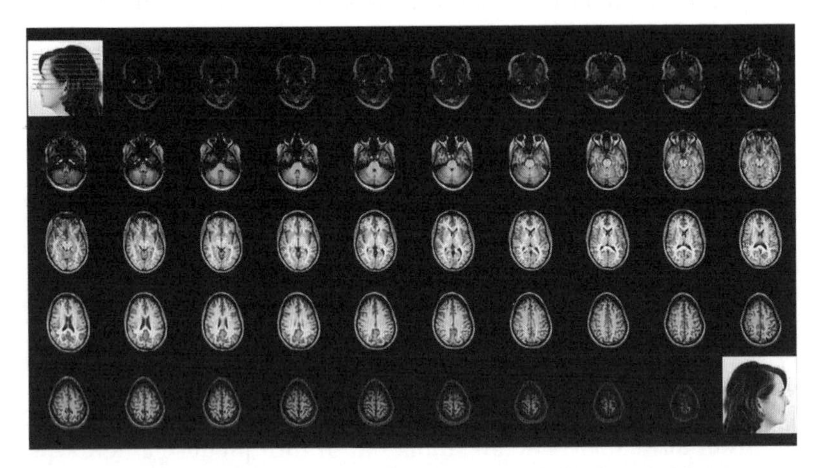

Figure 1.6 Portrait: Self-portrait while drawing, 2005. © Marta de Menezes

central to the belief that 'industrial and postindustrial technosciences ... implies the meticulous programming, of beautiful images'. In fact, these images are not only beautiful, but 'too beautiful', not as an indeterminate sentiment but rather the result of 'the infinite realization of the sciences, technology and capitalism' (Lyotard, 1993: 122). Artworks such as de Menezes's can also be seen as critical deconstructive practices since 'metaphysical complicity' cannot be given up without also giving up the critique of the complicity that is being argued against (Derrida, 1978: 281). These performances, whilst apparently complicit with dominant means of digital representation, attempt at the same time to destabilize those dominant structures by focusing on areas of concern relating to the commodification and consumerism of such technology.

Critical Art Ensemble (CAE.), through their 'recombinant theatre', have made technology, wetware and transgenics the focus of their work. For CAE, recombination 'typically denotes esoterica pertinent to molecular biology'. Although originally working with multimedia, CAE over recent years have concentrated on responding to the debates surrounding biotechnology. As 'tactical' mediaists the group have presented various interactive performance projects. For CAE, tacticality includes a willingness to be amateurs, to try anything, and to resist specialization. CAE's performance *Free Range Grain* (2004) added another unplanned dimension to their intention of creating a 'theatre of everyday life', in as much as they have found themselves in the midst of an aggressive investigation launched under the US bioterrorism laws. In reality what was found during a search of a member's property was equipment and materials that were to be used in the performance/exhibition It is still unclear what charges if any will ultimately be laid as a result of a certain State implemented paranoia. However, what is certain in this instance is that CAE have blurred the distinction between performance and everyday life and in keeping with all their projects have also endeavoured to open up dubious government practices to public scrutiny.

In conclusion, whilst the central distinctive aesthetic trait of digital practices is the utilization of the latest digital technology, the digital as a discourse cannot convert phenomena directly but depends on a preceding production of meaning by the non-digital. For instance, the avatar in *Blue Bloodshot Flowers* emulates the graphic design and animation of a recognizable representation, which is in this case a human head. The digital, like all formal systems, has no inherent semantics unless one is added. One must add meaning. Thus digitally processed contents require different than ordinary habits of reading – reading digital contents demands thinking in terms of 'indifferent differentiation'. A thinking that makes little distinction between the referent and meaning or for that matter between 'reality' and representation.[5]

It is my belief that 'digital' practices as experimental artworks and performances both serve as critique and have an *indirect* affect on the social and political, though a redefinition of this term is certainly needed, in as much as they question the very nature of our accepted ideas and belief systems regarding new technologies. In this sense, the digital does what all avant-garde art does; it is an experimental extension of the socio-political and culture of a milieu.

Finally, although much interest is directed towards new technologies, it is my belief that technology's most important contribution to art is the enhancement and reconfiguration of an aesthetic creative potential that consists of interacting with, and reacting to, a physical body. For, it is within these tension-filled (liminal) spaces of the physical and virtual interface that opportunities arise for new experimental forms and practice.

Notes

1 'Liminality', from *limen* (Latin: literally threshold) is a term most notably linked to Victor Turner who writes of a no-man's-land betwixt-and-between, a site of a 'fructile chaos ... a storehouse of possibilities, not by any means a random assemblage but a striving after new forms' (Turner, 1990: 11–12). My own use of the term includes certain aesthetic features described by Turner, but emphasizes the corporeal, technological and chthonic (Greek: back to the earth) or primordial. Other quintessential features are heterogeneity, the experimental and the marginalized. Therefore, *liminal* performance can be described as being located at the edge of what is possible (Broadhurst, 1999a: 12).

2 'The term alone suggests a field of study that is pregnant and full of promise. It is a large field of study uniting concepts and techniques from many disciplines ... At the heart of cognitive neuroscience ... lies the fundamental question of knowledge and its representation by the brain ... Cognitive neuroscience is thus a science of information processing ... one can identify key experimental questions ... How is information acquired (sensation), interpreted to confer meaning (perception and recognition), stored or modified (learning and memory) ... and to communicate (language)?' (Albrecht and Neville, 2001: li).

3 Open Sound Control was created by the Center for New Media and Audio Technologies (CNMAT) at the University of California, Berkeley in the 1990s.

4 A significatory practice which involves such non-linguistic modes as those provided by the semiotics of corporeal gesture. See (Broadhurst, 1999a, 1999b, 2002, 2004a, 2004b).

5 For a more detailed discussion of the concepts of 'differentiation' and 'de differentiation', see Scott Lash (1990: 5–15). See also Broadhurst (1999a: 177).

References

Albrecht, Thomas D. and Helen J. Neville. (2001). 'Neurosciences', in *The MIT Encyclopedia of The Cognitive Sciences*. Ed. Robert Wilson and Frank Keil. Cambridge, MA: The MIT Press, pp. li–lxxii.

Bogue, Ronald. (1989). *Deleuze and Guattari*. London and New York: Routledge.

Broadhurst, Susan. (1999a). *Liminal Acts: A Critical Overview of Contemporary Performance and Theory*. London: Cassell; New York: Continuum.

——. (1999b). 'The (Im)mediate Body: A Transvaluation of Corporeality', *Body & Society* 5.1 (March): 17–29.

——. dir. (2001). *Blue Bloodshot Flowers*. Performer Elodie Berland. Music by David Bessell. Technology provided by Richard Bowden, University of Surrey. Brunel University (June); The 291Gallery, London (August).

——. (2002). 'Blue Bloodshot Flowers: Interaction, Reaction and Performance', *Digital Creativity* 13.3: 157–63.

——. (2004a). 'Interaction, Reaction and Performance: The Jeremiah Project', *The Drama Review* (MIT Press) 48.4 (Winter): 47–57

——. (2004b). 'Liminal Spaces', in *Mapping the Threshold: Essays in Liminal Analysis* (Studies in Liminality and Literature 4). Ed. Nancy Bredendick. Madrid: Gateway Press, 57–73.

Coniglio, Mark and Dawn Stoppiello. (2005) Troika Ranch Website. <www.troikaranch.org/>, accessed 20 June 2009.

Crick, Francis. 1994. *The Astonishing Hypothesis*. London: Simon & Schuster Ltd.

Crick, Francis and Christof Koch. (1990). 'Towards a Neurobiological Theory of Consciousness', *Seminars in the Neurosciences* 2: 263–75.

Critical Art Ensemble, Beatriz da Costa, and Shyh-shiun Shyu. (2004). 'Free Range Grain'. <http://www.critical-art.net/biotech/free/index.html>, accessed 26 May 2009.

Cunningham, Merce, chor. (2000). *Biped*. Computer-enhanced graphics: Paul Kaiser and Shelley Eshkar. Music: Gavin Bryars. Costume Designer: Suzanne Gallo. Barbican Centre, London, 11 October (Premiered at Zellerbach Hall, Berkeley, California, 23 April 1999).

Deleuze, Gilles. (2003). *Francis Bacon: The logic of sensation*. Trans. Daniel W. Smith. London and New York: Continuum.

Deleuze, Gilles, and Félix Guattari. (1999). *A Thousand Plateaus: Capitalism and Schizophrenia*. Trans. Brian Massumi. London: Athlone.

De Menezes, Marta. (2000). *Nature?* Developed at the Institute of Evolutionary and Ecological Sciences, Leiden University, The Netherlands and exhibited at the 'Next Sex: Sex in the Age of Its Procreative Superfluousness', Ars Electronica, Linz, Austria.

——. 2002. 'Functional Portraits', Marta de Menezes's website <http://www.martademenezes.com/>, accessed 20 June 2009.

Derrida, Jacques. (1978). 'Structure, Sign and Play in the Discourse of the Human Sciences', in *Writing and Difference*. Trans. Alan Bass. Chicago: University of Chicago Press, pp. 278–93.

Dowling, Peter, Robert Wechsler and Frieder Weiss. (2004). 'EyeCon – a motion sensing tool for creating interactive dance, music and video projections'. Proceedings of the Society for the Study of Artificial Intelligence and the

Simulation of Behavior (SSAISB)'s convention: Motion, Emotion and Cognition at University of Leeds, England, (March): 1–7.

Lash, Scott. (1990). *Sociology of Postmodernism*. London: Routledge Press.

Lyotard, Jean-François. (1988). 'An Interview with Jean-François Lyotard'. Interviewed by Willem van Reijan and Dick Veerman, trans. Roy Boyne, in *Theory, Culture and Society* 5.2–3 (June): 277–309.

——. (1989). 'The Dream Work Does Not Think', in *The Lyotard Reader*. Ed. Andrew Benjamin, trans. Mary Lydon. Oxford: Basil Blackwell, pp. 19–55.

——. (1993). *The Inhuman: Reflections on Time*. Trans. Geoffrey Bennington and Rachel Bowlby. Cambridge: Polity Press.

Merleau-Ponty, Maurice. (1962). *Phenomenology of Perception*. Trans. Colin Smith. London: Routledge.

Ramachandran, V. S. and Sandra Blakeslee. (1999). *Phantoms In The Brain*. New York: Quill.

Ramachandran, V.S. and Edward M. Hubbard. (2001). 'Synaesthesia: A Window into Perception, Thought and Language', *Journal of Consciousness Studies* 8: 3–34.

Stelarc. (2002). 'Towards a Compliant Coupling: Pneumatic Projects, 1998-2001', in *The Cyborg Experiments: The Extensions of the Body in the Media Age*. Ed. Joanna Zylinska and Gary Hall. London/New York: Continuum, pp. 73–8.

——. (2003). *Muscle Machine* Performance Premiere. A collaboration involving the Digital Research Unit, Nottingham Trent University and the Evolutionary and Adaptive Systems Group, COGS at Surrey University. 291 Gallery, East London, 1 July.

——. (2004). 'Stelarc 1/4 Scale Ear', Stelarc's website <http://www.stelarc.va.com.au/extra_ear/index.htm>, accessed 20 November 2004.

Troika Ranch. (1996). *The Electronic Disturbance*. Created by Mark Coniglio and Dawn Stoppiello. Performed by Dawn Stoppiello, Gail Giovaniello, Lana Halvorsen, Rose Marie Hegenbart, Ernie Lafky and Joan La Barbara. The Kitchen, New York (April).

——. (2003). *The Future of Memory*. Created by Mark Coniglio and Dawn Stoppiello. Performed by Dawn Stoppiello, Danielle Goldman, Michou Szabo, and Sandra Tillett. Premiered at The Duke, 42nd Street, New York (February).

——. (2004). *Surfacing*. Choreographer: Dawn Stoppiello. Music & Video: Mark Coniglio. Costume Design: Wendy Winters. Lighting Design: Susan Hamburger. Set Design: David Judelson. Performed by Danielle Goldman, Patrick Mueller, Michou Szaabo and Sandra Tillett. Premiered at Danspace Project, New York (May). I attended a performance at Chancellor Hall, Essex, 12 May 2005.

——. (2006). *16 [R]evolutions*. Choreographer: Dawn Stoppiello. Music and Video: Mark Coniglio. Set Desigh: Joel Sherry. Lighting: Susan Hamburger. Performers: Robert Clark, Johanna Levy, Daniel Suominen and Lucia Tong. New York premiere Eyebeam Arts and Technology Center, Chelsea, January.

Turner, Victor. (1990). 'Are there universals of performance in myth, ritual, and drama?', in *By Means of Performance*. Ed. Richard Schechner and Willa Appel. Cambridge: Cambridge University Press, pp. 1–18.

Zeki, Semir. (1999). *Inner Vision: An Exploration of Art and the Brain*. Oxford: Oxford University Press.

Zeman, Adam. (2002). *Consciousness: A User's Guide*. London: Yale University Press.

2
Texts from the Body

Tracey Warr

Introduction

A good third of our psychic life consists in ... rapid premonitory prespective views of schemes of thought not yet articulate.

(James, 1950: 253)

Bruce Gilchrist's artworks are a sustained investigation of what psychologist William James termed, as cited above, 'thought not yet articulate'. Gilchrist combines poetic strategies with customised brain imaging and biofeedback technologies to draw 'texts' from the mute body and to image and manifest these texts for audiences. His artworks journey in to the unconscious unlanguaged parts of consciousness, into prehension and preconsciousness.

In their study of the neuroscience of free will, Bernard Libet, Anthony Freeman and Keith Sutherland write that 'the development of non-invasive imaging techniques ... may soon be able to provide a much better handle on subjective experience' (1999: xxii). They point out, however, that PET and functional MRI can only tell us *where* in the brain a change in neural activity is occurring. Brain imaging technologies do not tell us *what* is going on:

> As far back as Leibniz it was pointed out that if one looked into the brain with a full knowledge of its physical makeup and nerve cell activities, one would see nothing that describes subjective experience ... externally observable and manipulable brain processes

> and the related subjective introspective experiences must be studied simultaneously ... to understand their relationship.
>
> (Libet, 1999: 55)

Gilchrist's artworks turn physiological data into something resembling notation. In a series of works exploring sleep in the late 1990s culminating in *Divided by Resistance* (1996), audience members could communicate with Gilchrist's sleeping body using an adapted Morse code.[1] He had learnt the code for a number of phrases which were 'administered' to his sleeping body by audience members through mild electrical signals:

> Can you respond to this question?
> x x – – x x – –
>
> I'm not certain what I'm interfacing with
> x x – – x x – – x
>
> Am I in danger of being misunderstood?
> x x x – – x x – –
>
> (Artemergent, 2009)

The artist's sleeping body 'responded' to the audience members' questions and observations through amplified body sounds. In *Thought Conductor* (1997–2001) Gilchrist used the EEG (electroencephalograph) of composers in the act of music composition, combined with the EEG of audience members, to generate a musical score played during live performances. In *Spacebaby* (2008) blood samples taken from Gilchrist and his collaborator Jo Joelson were converted into gene expression imagery in the form of chroma and luma data.[2] Using this mixture of techniques from biological science, music, computer science and performance art, Gilchrist has generated a series of body texts that nevertheless remain tantalisingly indecipherable. The following account of Gilchrist's work discusses his explorations of twenty-first-century body texts in relation to James's nineteenth-century account of the stream of consciousness.

Divided by resistance: thinking fluid

> Countless layers of ideas, images, feelings, have fallen successively on your brain as softly as light. It seems that each buries the preceding, but none has really perished.
>
> (Baudelaire, cited in Gilchrist, 1995: n.p.)

Gilchrist's works enact Francisco J. Varela's proposition that the body is a portable laboratory. Varela describes Freud's development of psychoanalysis, the Wurzburg School of Introspectionism and Husserl and Merleau-Ponty's phenomenology as versions of the self-laboratory (1999: n.p.). Gilchrist explores the body engaged in a fluid process of thinking without language. He explores what is in between the languaged and the super-ephemerality of consciousness:

> What it means for any thought to be sub- or pre- conscious has yet to be resolved ... Are sub- or pre- conscious thoughts completely unacknowledged or are they just infinitesimally fleeting, communicating in an ellipsis of a very few words or a partial image, our knowledge of them obscured by the reactions they trigger – one dim moment of consciousness lost in the glare of the next few brighter ones?
>
> (Bricklin, 1999: 89)

> Reality proves to be more a no-thing-ness than a some-thing-ness, and about as substantial as a passing cloud.
>
> (Gilchrist, 1995: n.p.)

In *The Principles of Psychology* (first published in 1890) James discusses what he terms 'transitive' thought. I am quoting here from James's description at some length, because his writing employs a heightened capability for introspection which is illuminating for an understanding of Gilchrist's body texts:

> [Consciousness is] continuous ... without breach, crack, or division ... Consciousness, then, does not appear to itself chopped up in bits. Such words as 'chain' or 'train' do not describe it fitly as it presents itself in the first instance. It is nothing jointed; it flows. A 'river' or a 'stream' are the metaphors by which it is most naturally described. *In talking of it hereafter, let us call it the stream of thought, of consciousness, or of subjective life* ... what strikes us first is [the] different pace of its parts. Like a bird's life, it seems to be made of an alternation of flights and perchings ... The resting-places are usually occupied by sensorial imaginations of some sort, whose peculiarity is that they can be held before the mind for an indefinite time, and contemplated without changing; the places of flight are filled with thoughts of relations, static or dynamic, that for the most part obtain between the matters contemplated in the periods of comparative rest.
>
> *Let us call the resting-places the 'substantive parts,' and the places of flight the 'transitive parts,' of the stream of thought.* It then appears that

> the main end of our thinking is at all times the attainment of some other substantive part than the one from which we have just been dislodged. And we may say that the main use of the transitive parts is to lead us from one substantive conclusion to another.
>
> (1950: 237–43)

James uses metaphors of a flowing stream and a bird's flight to try to describe transitive thought. He writes that no existing language is capable of doing justice to it. Psychologists, he writes, are challenged 'to *produce* these transitive states of consciousness' when that would be akin to asking 'in what place an arrow is when it moves':

> Now it is very difficult, introspectively, to see the transitive parts for what they really are. If they are but flights to a conclusion, stopping them to look at them before the conclusion is reached is really annihilating them. Whilst if we wait till the conclusion *be* reached, it so exceeds them in vigour and stability that it quite eclipses and swallows them up in its glare. Let anyone try to cut a thought across in the middle and get a look at its section, and he will see how difficult the introspective observation of the transitive tracts is. The rush of the thought is so headlong that it almost always brings us up at the conclusion before we can arrest it. Or if our purpose is nimble enough and we do arrest it, it ceases forthwith to be itself.
>
> (James, 1950: 243–4)

James compares attempting to capture the fluid continuous rapidity of ephemeral consciousness to trying to turn the gas up fast enough in order to see darkness or trying to catch a snowflake when it melts on contact with the warm palm of the hand. Gilchrist's body texts are like this. Something is there manifested but it remains just beyond reach, not quite graspable or decipherable:

> We ought to say a feeling of *and*, a feeling of *if*, a feeling of *but*, and a feeling of *by*, quite as readily as we say a feeling of *blue* or a feeling of *cold*. Yet we do not: so inveterate has our habit become of recognizing the existence of the substantive parts alone, that language almost refuses to lend itself to any other use.
>
> (James, 1950: 245–6)

In a more recent study neuroscientist Antonio Damasio has described this transitive thought as 'the feeling of what happens' (1999). Damasio argues that thinking is done by means of patterns of nerve cell activation produced in response to sensory perceptions of the external world on the one hand, and monitoring of the internal world of body states and emotions on the other. Cognitive representations of the external world interact with cognitive representations of the internal world. Perceptions interact with emotions and consciousness is constantly drawing on both types of information, external and internal, and using them to adapt behaviour (Charlton, 2000: 99–101).[3]

The title of *Divided by Resistance* suggests the solipsism of the individual never able to fully communicate across the divide from one subjectivity to another. Gilchrist's works function through complex layers of body in performance (artists and audience members), installation, biofeedback technologies and physiological data, computer databases and software. In *Divided by Resistance* Gilchrist's sleeping body lies in the centre of the space hooked up to various equipment. Electrodes on his head and body connect him to an EEG machine. His EEG is projected onto the wall behind him and fed into a computer search engine looking for matches in its dream database. Infra-red sensors worn over the eyes indicate when he is in REM (rapid eye movement which indicates dreaming). The artist's brainwave signals are converted to voltage that drives small vibrating pads in a chair audience members can sit in, and in a jacket and shoes that they can wear. The 'resistance' of the voltage attempts communication across the divide. The communication is not something we can recognize or articulate as a known language. The audience, gently pummelled by Gilchrist's sleeping brainwaves, wearing his dreaming thoughts, are experiencing an unknown language. The technologies create an imperfect communication conduit between sleeper and individual audience members.

Gilchrist undertook a year of sleep research in his studio. He recorded his dreaming EEG patterns. When his computer detected that he was in mid-REM it woke him up and reminded him to make a written record of what he could remember of his dreams:

> It has been observed that it is much easier to remember, or to 'sustain', one's dream upon waking, if the awakened dreamer does not move his or her body in any way. If one rolls over, or simply extends an arm, the dream spills away with the gesture.
>
> (Mary Warnock, qtd in Scarry, 1985: 354)

Gilchrist created simulations of his dreams as Quicktime movies. Collaborator Jonny Bradley wrote database and search software (The

DreamEngine) that could match live EEG patterns with stored EEG patterns and their associated Quicktime files of dreams. At the ICA The DreamEngine probed Gilchrist's sleeping body and used it as a switch to turn on and off projections of these dream movies. A complex array of sensorial translations are going on in this work, as the sleeping artist and his biomonitor outputs are converted to images, sounds and tactile experiences for the audience. In Gilchrist's database aesthetic 'creative records are retrieved and presented in the manner of a search engine but with the engine responding to an attenuated sensuousness' (Artemergent, 2009).

Gilchrist writes:

> I was interested in the way external signals become symbolically incorporated into dream content ... Detached from our critical faculties we are passive witnesses to the most extraordinary inventions of the under-mind.
>
> (Gilchrist and Warr, 2000: n.p.)

Gilchrist's body language is indexical as opposed to representational (see Krauss, 1985). It employs the body directly to make a mark rather than a secondary representation. His sleeping body performances are primary, like the handprint, the footprint, the body fluid smear or the body imprint.

Gilchrist's texts and messages from the sleeping body convey a sense of an inarticulate language. But a language that is a contingent flux rather than something that can be represented through atomistic discrete units and taxonomic representational understanding. 'Language is a hierarchical combination of bits. Liquid ... is indivisible' (Bois and Krauss, 1997: 124). Many twentieth-century novelists, including James Joyce, Virginia Woolf, Dorothy Richardson and Clarice Lispector, tried to write James's 'stream of consciousness', but a true stream of consciousness could not be captured in language. Once enunciated, it becomes something else – something concrete, defined, an object rather than a formless flow. These novelists had to stage the stream of consciousness.

Georges Bataille's notion of 'the informe' is useful here. In his 'Critical Dictionary' he describes 'the informe' as performing an operation of declassification and disordering of taxonomy (1929: 382). It is, he says, performative, like an obscene word or a languageless grunt, hrumph, hah. It resembles a wordless voicing, like throat singing, a cough or a sneeze, an orgasm or a birth ululation. Bataille's 'Critical Dictionary' is a disordered series of non-definitions of arbitrary words and phrases, a volatile taxonomy allowing categorical ruptures and undermining the materiality and stability of text. It is an act of sabotage against a disembodied academia. Bataille joins lexical units from the symbolic code with extralinguistic practices that are charged with libidinal intensity. Employing Dada tactics, he makes text of ink spots and belches (see Bois and Krauss, 1997).

A series of other works preceded and evolved out of *Divided by Resistance*. In *Ways of Asking for Reasons* (1994) Gilchrist and a group of other sleepers inverted their sleep patterns and slept on a shale heap exploring the potential of group dreaming influenced by their environment.[4] In *Transmutations* (1996) at the Zap Club in Brighton, Gilchrist's thumb print was tattooed onto his shoulder and his body response to the tattooing was amplified and 'experienced' by audience members by means of GSR (a Galvanic Skin Response Unit). Gilchrist also used GSR in a sleep performance at *The Incident* in Fribourg, Switzerland (1995) to amplify his sleeping body sounds into a space. In *Sonic Bloom* (1998) Gilchrist wired up plants and converted their signals into audible sound. (See Artemergent, 2009; Warr, 1996; and Walwin, 1997 for further descriptions and discussion of Gilchrist's sleep works).

Chairs and databases are recurring motifs in Gilchrist's work and in his collaborations with Jo Joelson. The 'seat of consciousness' chair in *Divided by Resistance*, designed by Juggernaut, was the first in a series of chairs providing audience body interfaces. In London Fieldworks' *Gastarbyter* (1998) audience members experienced Dugal McKinnon's composition of the sound and vibrations of the city through a chair. In *Polaria* (2002) audience members sat in a transparent chair with their hands on metal plates that enabled their physiological data to control the lighting installation (Gilchrist and Joelson, 2002). *Syzygy* (1999) mixed physiological and environmental data and conveyed it via a database to a responsive smart sculpture (See London Fieldworks, 2009, and Gilchrist and Joelson, 2002).

The works 'reference biometrics, brain machine interfaces (PET etc) and the burgeoning role of databases across all conceivable sectors' (Artemergent, 2009). Databases used by commercial companies, government, police and other authorities are relentlessly collecting and exploiting information on us. Our personal human data is interpreted and employed by remote inhuman IT and statistical interfaces which frequently throw up both ludicrous and unjust outcomes. Commercial exploitation of our data attempts to corner us in an inescapable labyrinth of consumption.

Screen avatars, smart border biometrics, cosmetic surgery, the human genome project, genetic modification, a borderless and id-less cyberspace

– the twenty-first-century body is enmeshed in contradictions of fixity and flux, authenticity and simulation. In January 2004, Italian philosopher Giorgio Agamben refused to give a lecture in the United States because he would have been required to give up his biometric information to enter the country. Agamben argues that contemporary society is tending toward data collection and use procedures that criminalize all citizens, placing the population under permanent suspicion and surveillance (Agamben, 2004).

Thought Conductor and *Thought Pavilion*: bio-music and bio-sculpture

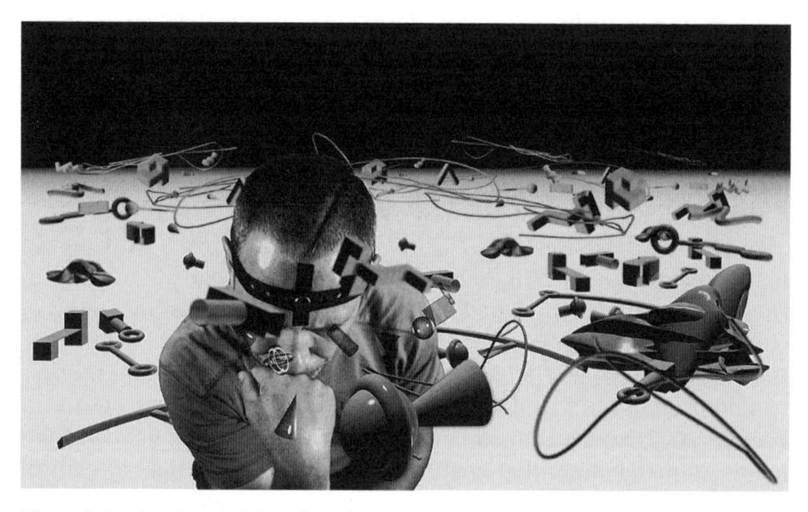

Figure 2.1 London Fieldworks, *Thought Pavilion*, collage

The first version of Gilchrist and Bradley's *Thought Conductor* was performed in November 1997 at The Place in London. Nikki Yeoh composed a score and sat thinking through it while she was connected to a digital biomonitor. Her EEG was projected onto the wall and software translated this into musical notation played live by Piano Circus on six Steinway grand pianos.

The second version presented in Stavanger Konserthus in June 2000 incorporated Bradley's database software created for *Divided by Resistance*. The EEG of 12 composers was recorded whilst they wrote compositions for a string quartet. The files of their EEG were associated with midi versions of their compositions and archived in a database. During the performance four audience members were connected to a biomonitor. They were a bricklayer, a jeweller, a yogi and a painter. The string quartet then played the combined scores generated by the composers' and the audience members' brainwaves. In a version presented in Holywell Music Rooms in Oxford in October 2001 as part of the *OX1* festival, the audience members contributing their brain waves to the score were an origami enthusiast, an accountant, a journalist and an academic. Mariam Fraser argues, however, that *Thought Conductor,* 'is not ... a techno-scientific portrait that seeks to capture the "inner kinetic melody" of the individual who sits at the centre of the stage' (2005: 173).

In *Thought Conductor,* who is the conductor and whose thought is manifested in the performances? Gilchrist mixes 'owned' thought to such an extent that the performance approaches James's idea of an ownerless thought in the room:

> My thought belongs with my other thoughts, and your thought with your other thoughts. Whether anywhere in the room there be a mere thought, which is nobody's thought, we have no means of ascertaining, for we have no experience of its like.
>
> (James, 1950: 226)

Fraser argues that in *Thought Conductor,* 'the entire event listens to itself' (2005: 179).

Gilchrist's work has been highly iterative with one work developing from a previous work in a sustained meditation on his key themes. Developing out of *Thought Conductor* Gilchrist has worked with Joelson to evolve their current proposal for *Thought Pavilion.*[5] *Thought Pavilion* aims to create thought sculptures through the combination of a database approach and manufacturing techniques, using rapid-prototyping technology and 3D printers which will render the shapes of thought as solid physical objects. Gilchrist and Bradley have so far compiled a database (entitled 'The Perception Depository') of the brainwaves of 300 volunteers at the point of discerning primitive shapes embedded in autostereographic imagery. In this work the subtle and ephemeral will be translated into concrete building blocks. Gilchrist has described his work as 'the objectification of phenomena associated with cognitive processes within a poetic application of technology' (Gilchrist and Warr, 2000: n.p.).

Gilchrist asks:

> Can a participant [in *Thought Pavilion*] claim intellectual copyright to brainwaves captured and any resulting 'artworks'? In the light of the ethical and moral issues surrounding the Human Genome Project, could the potential creative application of the contents of

> The Perception Depository be considered a public work of art or an artwork disinterred from the public?
>
> (Gilchrist and Warr, 2000: n.p.)

A collaborative approach, with other artists, scientists, audience members and environments is a characteristic of Gilchrist's work. His work raises issues around individuality, collectivity and authorship. Fraser points out the denial of authorial locus in *Thought Conductor*, where 'the detour through the digital ensures that the break between the interiority of the composer and the score ... is maintained' (2005: 180).

Gilchrist's work is concerned with the dynamic and co-creative interactions of embodied consciousness with the environment it is immersed in:

> There can be no single author of contingency and no single subject-listener ... the identity of the event, in short, is defined not by any one of its (individual) components (such as the author-composer), or even by the sum of its components (all that a musical performance involves). It lies, rather, in the singular *becoming-together* of reciprocal prehensions.
>
> (Fraser, 2005: 179)

Figure 2.2 London Fieldworks, *Spacebaby*, Roundhouse, London 2006, in the *Space Soon* event presented by Arts Catalyst. Photo: Kristian Buus

Conclusion

The sources of Gilchrist's collaboration with Bradley, and with Joelson as London Fieldworks, range through contemporary science and technologies to science fiction, the history of science, Victorian science and instrumentation and the work of esoteric writers and practitioners.[6]

Part of Gilchrist's research methodology as an artist is to submit himself as a lab rat to scientific researchers. Developing the sleep works that culminated in *Divided by Resistance*, for instance, he volunteered at the Clinical Psychology Sleep-Research Laboratory at the University of Texas at Austin and drew on the work of sleep researchers including Stephen La Berge and Keith Hearne. He volunteered as a PET scan subject in order to develop his own understanding of current brain and body imaging technologies. In *Spacebaby* (Figure 2.2) this lab rat approach continues into the work itself where Gilchrist and Joelson slept in hibernation tubes and scientists extracted regular blood samples from the sleeping artists to produce gene expression data. *Spacebaby* draws on scientific research into suspended animation, hibernation, circadian rhythms and gene expression.[7]

The overriding concern in all of Gilchrist's work is to make the body 'speak'. 'Suppose we try to recall a forgotten name', writes James:

> The state of our consciousness is peculiar. There is a gap therein; but no mere gap. It is a gap that is intensely active. A sort of wraith of the name is in it, beckoning us in a given direction, making us at moments tingle with the sense of our closeness, and then letting us sink back without the longed-for term. If wrong names are proposed to us, this singularly definite gap acts immediately so as to negate them. They do not fit into its mould. And the gap of one word does not feel like the gap of another ... When I vainly try to recall the name of Spalding, my consciousness is far removed from what it is when I vainly try to recall the name of Bowles ... The rhythm of a lost word may be there without a sound to clothe it; or the evanescent sense of something which is the initial vowel or consonant may mock us fitfully, without growing more distinct.
>
> (1950: 251–2)

Black marks on a piece of paper symbolize sounds, movements, pace, silences, pauses, thinking. Chess moves can be represented in notation, as can the pitch and duration of sound or dance movements. A score is a physical readable representation of music, of sound. Text is marks on paper representing language but also appearing conjured in the mind of the reader. The script is a representation of spoken words in a play, a film or a broadcast. Texts, scores, scripts, notations translate the immaterial and the somatic into the material. They transfer bodily, sensory experiences to paper and then are recreated from that paper representation in the mind or actions of another reader (from composer to musician, from writer to reader, from Gilchrist's body to yours).

Gilchrist's body 'texts' are indecipherable. With considerable scientific knowledge we might be able to make something of the EEG patterns shifting like lightning jags or of the coloured blizzard of Gene Expression Data we are presented with. But that is not the point of this work. Instead we are presented with texts which we *know* but cannot decode.

Roland Barthes's contends that in writing:

> Everything is to be *disentangled,* nothing *deciphered* ... the space of writing is to be ranged over, not pierced; writing ceaselessly posits meaning ceaselessly to evaporate it, carrying out a systematic exemption of meaning.
>
> (1977: 147)

In an unpublished essay on Barthes's idea of 'style' in *Writing Degree Zero* (1967), Alan Reed comments that style:

> Draws something from beyond language, and as such efforts to decipher it lead to the frustration of trying to speak the unspeakable. And yet, insofar as this unspeakable thing manifests in the text, it speaks. It does not say itself, but it does say something.
>
> (2008, n.p.)

Gilchrist's body texts say themselves and they say something. What it is, is on the tips of our tongues.

Acknowledgements

I would like to thank Bruce Gilchrist, Jo Joelson and Jonny Bradley for conversations about the works discussed above. Thanks to London Fieldworks for permission to reproduce the images. I would also like to thank Alan Reed for permission to quote from his unpublished essay, and the Something Like Spit collective, who contributed to my protracted engagement with the writings of William James.

Notes

1 *Divided by Resistance* won the ICA/Toshiba Art & Innovation Award.
2 Working in collaboration with Jo Joelson, Gilchrist appears as London Fieldworks (see London Fieldworks, 2009). Many of London Fieldworks'

projects also involve a further network of other artistic and scientific collaborators.

3 See Bruce Charlton for a useful 'lay' summary of Damasio's work (Charlton, 2000: 99–101).

4 In 1974 Susan Hiller's *Dream Mapping* explored drawing notation for dreams and group dreaming in a mushroom field in Hampshire.

5 *Thought Pavilion* is a proposed public art work for Oldham. Its development has been supported through Gilchrist's Arts and Humanities Research Board Research Fellowship 2002–05 based at Oxford Brookes University, Joelson's Research Fellowships at Leicester University and South Bank University, and through London Fieldworks' commission in 2005 from the London Science Museum's SMAP5 BIG IDEAS fund.

6 The title of Gilchrist's recent collaboration with Joelson (as London Fieldworks) *Spacebaby: Guinea Pigs Don't Dream* references the territory of science fiction, and especially Philip K. Dick's writing including *Do Androids Dream of Electric Sheep?* (1996). *Spacebaby* was a semi-fictional journey into genetic space. It was first presented as a live performance in *Space Soon* organized by Arts Catalyst at The Roundhouse in London in 2006 and then as a film shown at the Whitechapel Gallery in 2008.

7 Gilchrist and Joelson's contemporary scientific and technological collaborators have included genetic researchers at University of Leicester and rapid-prototyping engineers at Manchester Metropolitan University. Their work also draws on historical science such as Etienne-Jules Marey's chronophotographs of the movement of blood, Bela Julesz's work on pattern and depth perception at the Bell Laboratories in the 1950s and 1960s, and Christopher Tyler's work on autostereograms at the Smith-Kettlewell Institute in the 1970s and 1980s. Another source for Gilchrist's work has been the interface between art and esoteric thought, and especially Theosophy. In their study of auras and thought-forms, *Occult Chemistry* (1908), theosophists Annie Besant and Charles Leadbeater claimed to have perceived the constituent parts of chemical elements in advance of their scientific discovery. Theosophy had a significant impact on many artists and thinkers including Piet Mondrian, Wassily Kandinsky, Edvard Munch, Rudolf Steiner and Kasimir Malevich. Russian Theosophist P. D. Ouspensky's book *The Fourth Dimension: A Study of an Unfathomable Realm* (1909) and Kandinsky's *On the Spiritual in Art* (1911) also belong to the literature crossing art and theosophy in this period. Gilchrist's interest in thought forms relates to Kandinsky's study of his own synaesthesia, Aleksandr Scriabin's colour organ and Mikhail Matiushin's synaesthetic paintings such as *Red Ringing*. Matiushin was a member of the Studio of Spatial Realism in the Petrograd State Free Art Teaching Studios in the 1920s, where, along with the three Enders artists (two brothers and a sister), he researched methods of expanded viewing such as seeing with each eye independently, expanding peripheral vision and seeing through the back of the head (Tuchmann, 1986). Gilchrist undertook research in the library of the Theosophical Society as part of his development work for *Thought Pavilion*. In his performance *The Discarnate* (1996) in Glasgow, Gilchrist worked with an auric healer and a warlock from the Glasgow Theosophical Society exploring out of body experience as well as sleep. In the *Sutemos/ Twilight* exhibition at the Centre for Contemporary Art Vilnius (1998), Gilchrist used Lithuanian shamans as the subjects for his biofeedback performance.

References

Books, Articles, Websites

Agamben, G. (2004). 'No to Bio-Political Tattooing', *Le Monde*, 10 January, online at <http://www.truthout.org>, accessed 4 February 2009.

Artemergent (2009). <http://www.artemergent.org.uk>, accessed 4 February 2009.

Barthes, R. (1967). *Writing Degree Zero*. New York: Hill & Wang.

——. (1977). 'The Death of the Author', in *Image Music Text*. New York: Hill & Wang.

Bataille, G. (1929). 'A Critical Dictionary: The Informe', *Documents*, 1.7: 382. Reprinted in Stoekl, A. (ed.) (1985), *Georges Bataille: Visions of Excess: Selected Writings 1927–1939*. Minneapolis: University of Minnesota Press, pp. 31.

Bois, Y. and R. E. Krauss. (1997) *Formless: A User's Guide*. New York: Zone.

Bricklin, J. (1999). 'A Variety of Religious Experience: William James and the Non-Reality of Free Will', in *The Volitional Brain: Towards a Neuroscience of Free Will*. Ed. B. Libet, A. Freeman and K. Sutherland. Thorverton: Imprint Academic, pp. 77–98.

Charlton, B. G. (2000). 'Review of *The Feeling of What Happens*', *Journal of the Royal Society of Medicine*, 93: 99–101.

Damasio, A. (1999). *The Feeling of What Happens: Body, Emotion and the Making of Consciousness*. London: Heinemann.

Dick, P. K. (1996 [1968]). *Do Androids Dream of Electric Sheep?* New York: Ballantine.

Fraser, M. (2005). 'Making Music Matter', *Theory, Culture, Society*, 22.1: 173–89.

Gilchrist, B. (1995). 'Divided by Resistance: Dream Research and Neural Networks', unpublished essay, n.p.

Gilchrist, B. and T. Warr. (2000). 'Art as a First-Person Methodology in Consciousness Research', unpublished conference paper presented at *Toward a Science of Consciousness* at the University of Arizona in Tucson in April 2000. Abstract published in *Toward a Science of Consciousness: Tucson 2000* (Thorverton: Journal of Consciousness Studies), p. 162.

Gilchrist, B. and J. Joelson, J. (eds). (2002). *London Fieldworks: Syzygy/Polaria*. London: Black Dog Publishing.

James, W. (1950 [1890]). *The Principles of Psychology*, vol. 1. New York: Dover Publications.

Krauss, R. E. (1985). 'Notes on the Index: Parts 1 and 2', in *The Originality of the Avant-Garde and other Modernist Myths*. Cambridge, MA: MIT Press, pp. 196–209, 210–19.

Libet, B. (1999). 'Do We Have Free Will?', in Libet, Freeman and Sutherland, pp. 47–58.

Libet, B. A. Freeman and K. Sutherland (eds). (1999). *The Volitional Brain: Towards a Neuroscience of Free Will*. Thorverton: Imprint Academic.

London Fieldworks (2009). <http://www.londonfieldworks.com>, accessed 4 February 2009.

Reed, A. (2008). 'Barthes' Style', unpublished essay, np.

Scarry, E. (1985). *The Body in Pain: The Making and Unmaking of the World*. Oxford: Oxford University Press.

Something Like Spit. (2009). <http://somethinglikespit.org.uk>, accessed 4 February 2009.

Tuchmann, M. (ed.). (1986). *The Spiritual in Art*. Los Angeles: Los Angeles County Museum.
Varela, F. J. (1999). 'The Portable Laboratory', in *Laboratorium*. Ed. H. U. Obrist and B. Vanderlinden. Antwerp: Provincaal Museum voor Fotografie, n.p.
Walwin, J. (1997). *Low Tide: Writings on Artists' Collaborations*. London: Black Dog Publishing.
Warr, T. (1996). 'Sleeper: Risk and the Artist's Body', *Performance Research*, 1.20: 1–19.

Live Performances

Gilchrist, B. (1994). *Ways of Asking for Reasons*. Shale Heap, Cleveland. Part of *EarthWire*.
——. (1995). *Ways of Asking for Reasons*. Belluard Bollwerk, Fribourg, Switzerland. Part of *The Incident*.
——. (1996). *The Discarnate*. Centre for Contemporary Art, Glasgow.
——. (1996). *Transmutations*. Zap Club, Brighton.
——. (1998). *The Discarnate*. Centre for Contemporary Art, Vilnius, Lithuania. Part of *Twilight/Sutemos*.
——. (1998). *Sonic Bloom*. Garcia de Orta Botanic Gardens, Lisbon, Portugal. Part of *Expo '98*.
Gilchrist, B. and J. Bradley. (1996). *Divided by Resistance*. Institute of Contemporary Arts, London. Part of *Totally Wired*.
——. (2000). *Thought Conductor #2*. Konserthus, Stavanger, Norway.
——. (2001). *Thought Conductor #2.1*. Holywell Music Room, Oxford. Part of the *OX1: Oscillations and Vibrations Festival*.
Gilchrist, B., J. Bradley and J. Joelson. (1997). *Thought Conductor #1*, in *Winter Music*, The Place, London.
London Fieldworks. (1998). *Gastarbyter*, Institute of Contemporary Arts, London.
——. (1999). *Syzygy*. Institute of Contemporary Arts, London.
——. (2002). *Polaria*. The Wapping Hydraulic Power Station, London.
——. (2005). *Thought Pavilion*, proposal.
——. (2006) *Spacebaby*. The Roundhouse, London. Part of *Space Soon*.

3

De-Second-Naturing: Word Unbecoming Flesh in the Work of Bodies in Flight

Sara Giddens and Simon Jones

Since 1997, through a series of intermedial collaborations with musicians, video and sonic artists, Bodies in Flight have progressively interrogated the impact of digital technologies on our sense of our selves and our interrelationships with others, and how those technologies can be used in performance to expose this intimate process of incorporation into the human psyche – what Bodies in Flight call 'second-naturing'. This series of works has produced a sustained contemplation on contemporary human experience as *an interstices in-between* various discursive fields and their related technologies.

SIMON JONES: What I am describing are three crucial points in the development of Bodies in Flight's aesthetic investigation into the possibilities of performance, at which, working first without and then with(in) technology, came to address Heidegger's fundamental assertion about our so-called technological age – 'the essence of technology is by no means anything technological' (1953: 311). This is more of a reflection on past work than a manifesto for future endeavours; it attempts to understand *as possibilities* some of the consequences of the artistic decisions we took when encountering knowingly or unknowingly the technological, rather than account for them as intentions, as the minutiae of either making or interpreting the shows. It proposes that the technological offers us a privileged access into the general mechanisms whereby we all construct and sustain our everyday performances of our selves.

SARA GIDDENS: I remember being somewhat fearful of working with technology. Understanding technology as that industrial, analogue now digital beast resulting ultimately in all that heavy metal cluttering up the rehearsal rooms exacerbated by the need to hump it about and then wait for it to start up, break down and then start up again! Caught in between a genuine fear of not knowing (not being competent with) and feeling that everyone was 'doing it!' was the mid-1990s after all: and I was drawn to performance precisely because of its fleshy, visceral, communal liveness.

I'm not sure this quite deeply felt resistance has ever truly left me? Certainly as a result of my collaborative relationship with Simon, I came to change my focus (or re-focused), and began to appreciate that we have always worked with a range of technologies and that understanding and even respecting different rhythms (people and things) was very much part of the collaborative process that I love so dearly. I sensed there was an art to it – and this demanded patience!

Reassuringly I could perceive that the body was indeed a technology in amongst all those others, in fact I could see the body itself as an assemblage of very practical, and multifarious and wondrous technologies. I could hear the production of language itself as a technology, as an example of *the cleft cleaving* as we called it. The *natural* and the learnt technology, of the tongue and the co-ordination of its 16 muscles all having to work consensually and very hard, particularly in any Bodies in Flight show! Over time the rehearsal studio became a place to explore the different capacities of each technology, each element, and each separate line of flight and of course inevitably the gaps in between them.

Our attention to this absolute separation of elements became a resounding principle in our 1996 show *Do the Wild Thing!* (See *Flesh & Text* for archival material on all our works).

SJ: In order to begin to unpack the density of the basic eventness of performance, so that audience-spectators could disentangle themselves from its enveloping, white-hot interstices, we turned to Brecht. He had proposed in the 1920s a set of estrangements whereby the audience-spectators would be able to put themselves *as if* at a distance from the events unfolding on stage. By opening up the actual gaps both between and within media inherent in the theatrical experience itself, Brecht intended to open up the possibilities of engagement. This separation of the elements of performance was the basic aesthetic strategy we used in *Do the Wild Thing!* We later discovered that it had opened up the Pandora's Box of media in general, and the technologies that facilitated them, as a means of accessing the discontinuities of experience in general *as the basis for a performance of self in particular.*

SG: Here movement and speech, now flesh and text, were quite literally separated from one another. The man (who speaks almost all of

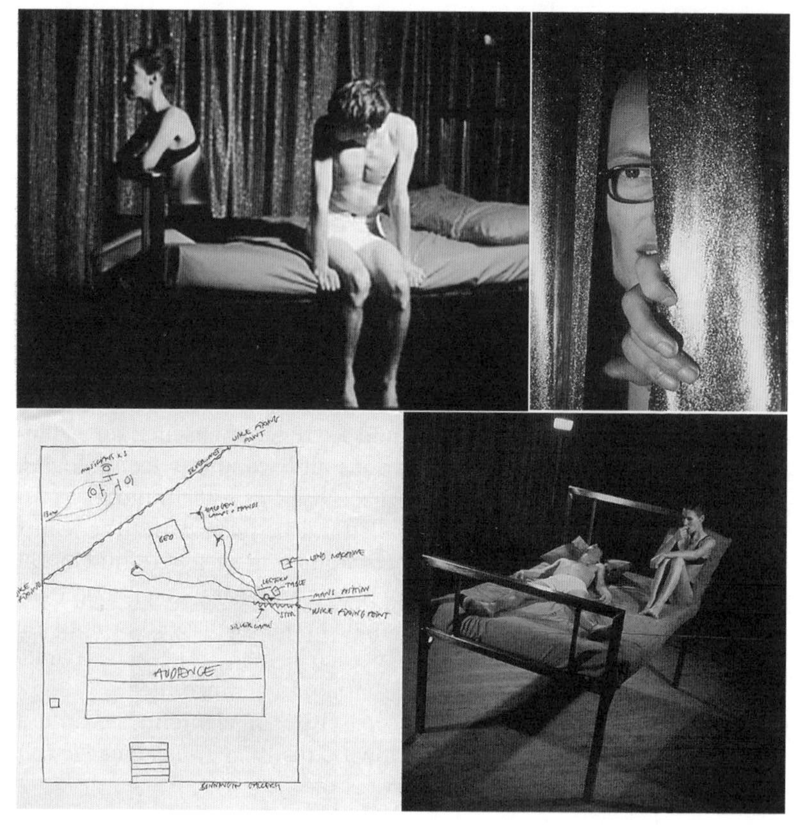

Figure 3.1 Shows and performers in order top left to bottom right: *Do the Wild Thing!* Performers Jane Devoy and Dan Elloway, Jon Carnall. Drawing by Bridget Mazzey, Jane Devoy and Dan Elloway. Photography: Edward Dimsdale

the text) is divided from the other two performers and from the audience by a curtain (Figure 3.1). This show was even rehearsed utterly apart – I was in one room with two performers (Jane Devoy and Dan Elloway) and a bed and Simon and Jon (Carnall) where in another with a chair and a microphone. In retrospect it was another great hunk of metal (the steel-framed, bespoke bed, the elongated proportions of which formed a platform, a kind of stage within a stage, designed by Bridget Mazzey) that provided a site upon which to really 'ground the movement'. It felt pointless to compete with the full and constant sound, the deliberate and at times oppressive tirade of words with constant movement. Less is more! We were learning to give space and time in order to allow the expression through and exploration of each technology. We had to find a different channel, a different register, to follow a gut instinct and employ our capacity to be able to focus down on the slightest nuance or gesture placed in amongst the quiet, the almost always still-ness and that somehow by focusing in on these, on the hush in the noise of the outpouring of words, we could render them important, special, gloriously unique.

In isolation, or separation, the movement and the words were able to afford, or offer up, many more possibilities. The interrelations between and in between the two were so much richer. We were not trying to illustrate one through (the technology of) another but, in placing the two side by side, these two could collude and collide by way of the third, the presence of the spectator with their attending. This process, a deconstruction of the performers' own movement, worked through and then given back to them so human, so vernacular, so personalized as to be inseparable or so it may seem, becoming seemingly second nature, a second-naturing. Choreographies made up of embodied practices – the turning of a head, the circling of an ankle, the curling of the hair. Thus the tiniest of details, the briefest of moments were homed in upon, fixed and then extended in space-time and made visible. We began to describe this process developed from the minutiae as *Micro-choreography*. It became a methodology: to make movement; break it down and build it up again. It's a technical procedure which creates time and space to focus in on, to attend to and dwell in around and alongside details of movement; and in doing so, this focusing reveals something somewhat hidden from our everyday view – 'The splinter in your eye is the best magnifying glass' (Adorno, 1978: 50).

Writing several years later, as we reflected on all of our work from 1989–2000 in a CD-Rom *Flesh and Text*, we came to understand more fully that what had been drawing our attention, fuelling our desire was this recombining of choreography and text to create opportunities for articulations '*when words move and flesh utters*'. Being at once fully and wholly in and in between these technologies allowed an opening up of space-time that revealed those rich and fascinating blind spots. I was where I wanted to be: here with these real, fleshy, sweaty, wondrous bodies amidst the sound of (t)his poetry: the image and the sound, seeing and hearing, a making visible and being heard – a duet.

SJ: This separation *hear—see* became the fundamental principle of Bodies in Flight's work: from it flowed not only a working method, but a series of philosophical enquiries into what it means to be human amidst the technological. 'The Man', who spoke behind the curtain, stood in for the spectator, seeing what they saw, saying what he saw, an interrogation of the scopic field of the scene and beyond, into the supposed motivations of the 'characters' before him. This function drives both performers and audience-spectators alike, but obviously from different perspectives. Furthermore, in saying what both he and they saw, he obliged the spectator to listen to what he made of that scene: in effect, he obliged a separating out of auditor and spectator within each participant. Through this pragmatics we found ourselves approaching the philosophical. Our separating echoed Foucault's commentary on Velasquez's *Las Meninas*:

> But the relation of language to painting is an infinite one. It is not that words are imperfect, or that, when confronted by the visible, they prove insuperably inadequate. Neither can be reduced to the other's terms: it is in vain that we say what we see; what we see never resides in what we say.
>
> (1970: 9)

Whilst the Man's investigation of the scene explores our contemporary culture's scopophilia, our fascination with appearance and the awkward persistence of beauty amidst the wasteland of consumerism and pornography, entering into the debates around gender and sexuality, its pervasive erotics posited a critical insight of phenomenologist Maurice Merleau-Ponty – *flesh as medium*. The discontinuity of living *between* these different media, or at least – *alongside* each other, is actually experienced *within* the one medium of our bodies:

> If the body is one sole body in its two phases, it incorporates into itself the whole of the sensible and with the same movement incorporates itself into a 'Sensible in itself'. ... It is this Visibility, this generality of the Sensible in itself, this anonymity innate to Myself that we have previously called flesh. ... The flesh is not matter, is not mind, is not substance. To designate it, we should need the old term 'element', in the sense that it was used to speak of water, air, earth and fire, that is, in the sense of a *general thing*, midway between the spatio-temporal individual and the idea, a sort of incarnate principle that brings a style of being wherever there is a fragment of being.
>
> (Merleau-Ponty, 1968: 138–9)

This 'incarnate principle' sums up Bodies in Flight's coupling of the performer with philosophical investigation, the body with text and with technology. In the event, the Man's dwelling upon the flesh became an endurance for both performer, as he reads unseen an hour-long monologue, and auditor-spectator, as they listen and look, slipping between sensory experiences and memories and fantasies, slipping in and out of concentration, the actual work of attending to the performance-event itself opening up the gaps between hearing and seeing, desires and perceptions.

SG: What is it you are able to do? How are you able to be? How are we together?

Perhaps it is not so strange that this separation led most clearly and deliberately to an attention to duets and duetting, and from there to the making of *Constants* (1998) and two different performers up close and very visible. This time the audience-spectators were seated individually in a broad sweep spiralling inwards, with *She*(ila) Gilbert and Patricia Breatnach moving in between them. *She* finally reaches the spiral's dead centre as Patricia exits with a final magnificent rush of youthful energy out of the furthest door. *Constants* was a show about memory and love, and, of course, that inevitable coupling – a loss of love.

The palpable presence of Sheila (then in her seventies) is my own overarching memory of this show: as I close my eyes I can still sense her. The frailty of a body, now more present than ever, a body beginning to fail, to fail to behave how its owner wanted it to behave, technically anyway. *She* used the walls and backs of chairs for support and her own physical 'capacities' became one structuring principle for the work itself (Figure 3.2). The other was the creation of the material through duets.

These duets with those who were extremely skilled in using more obvious technologies – video and audio artists Caroline Rye and Darren Bourne – became yet another layer in our emerging methodology, not so clearly differentiated in the final performance but central to the process. Each rehearsal focused upon one of the six possible permutations, employed roughly in this order in the actual show: movement and text, sound and text, sound and movement, movement and video, sound and video, text and video.

Applying this duetting principle throughout the making of *Constants* extended my concern for the minutiae that I had been drawn to in *Do the Wild Thing!* Using Caroline's hand-held cameras with built-in sound allowed me to explore the 'lost moment' (Blom and Tarin, 1989: 111), to accentuate together and in isolation both the sound of and the way the performers' bodies moved. These images were fed back

Figure 3.2 Shows and performers in order top left to bottom right: *Constants*. Performers: Patricia Breatnach and Shelia Gilbert. *DeliverUs*. Performers: Mark Adams and Polly Frame. Photography: Edward Dimsdale

through six video monitors and four speakers, arranged as part of the spiral alongside lighting, sound and video desks and their operators. Being in and alongside these bodies performing and acknowledging this very act, those digital technologies too had to be laid bare and be on full show. Yet the fleshy, sweaty, visceral, forgetful, awkward, magnificent loveliness of the performers and their audiences was even more heightened for me amidst that cacophony of metal gear. The more the audiences were drawn to the monitors, the more I was transfixed by the absolute beingness of the performers. Was this an abuse or simply a misuse of technology? Perhaps more obviously a reuse, as live material was recorded and played back in front of our very eyes and for all to hear, in real time, rewind and fast forward. An *aide-mémoire* butted up against our own memory of each individual event and alongside *She*'s memories. Certainly these duets allowed access to other depths of focus, other ways of attending to the technologies of the body, and perhaps most excitingly made possible a range of other manifestations of self. However, I found the aesthetics of man-made technology themselves irritating and often downright ugly. I realized I related to the monitors and speakers through their physicality. I found that their rhythms in the rehearsal room were so different to mine (both slower and quicker), that I felt interrupted by them.

SJ: If *Do the Wild thing!* began an aesthetic exploration through separation, followed by dwelling within the separated parts and persons of the scene, and *Constants* had actually introduced recognizably technological equipment into that scene (cameras, monitors, mixing desks), then *Deliver Us* (1999) intensified our interrogation of those different technological-textual fields. It took the word and its consort the move, within their discursive fields of language and choreography respectively, as the basic tools of a technological engagement which mirrored the more obviously inhuman – the machine-based use of video. In a specially constructed theatre within a theatre, the audience-spectators were seated around a small pit within which two performers appeared to sleep entwined. On each of the four sides of this pit was a television monitor, apparently with a blank screen. This scene, borrowed from Jean-Luc Godard's *Breathless* – the long scene of the lovers in bed after sex, talking and fooling about, provided the easily recognizable, everyday situation as access point for our continued aesthetic work.

For me, the Godard scene represented a holographic possibility within love: that one is somehow able to express the beginning and the end of love in each and every moment. What the physicist David Bohm might have called a 'holoworld' (1980: 134–51). Video-artist Caroline Rye carefully built up layers of mediation, gradually thickening the opacity of video as a medium with which to visualize the world, starting with a 'simple' live-relay of the action from camera to monitor, then introducing vision-mixing effects (freeze- and split-frame) to the 'live' image, before finally projecting 'life-sized' pre-recorded images of the performers, moving on to the performers actually moving, multiplying layers of images in different tempos as if one was suddenly watching a 'replay' of the whole history of the lovers' relationship.

Likewise the very speaking of the text presented the tongue with a *technological* problem: how to manipulate sounds that are essentially alien to it, since we have to *acquire* language; we have to learn *how to* speak. The know-how of speaking is a technological machine, a coupling of tongue and text. Just as the lovers feel that they have not fully recognized themselves until they have 'captured' their images on the monitors using the live-relay cameras, so their feelings cannot become fully real until they have spoken them. This need to capture and to phrase experience and our felt sense of it, and then our making sense of it, drives the lovers through the full potentiality of language as a tool. In a complicating similar to that of the video technology, language is layered through the work as three moods: the infinitive, a mood without person and time expressing passion's first flush; the simple present, a forensic mood when the lovers emerge from their ecstasies to interrogate their circumstances; and the subjunctive, a terrible mood, looking forward, connecting the lovers to all possible futures, when they turn from what they are to what they might become. Each new tense disorganizes their sense of selves as it paradoxically allows those selves self-expression: their love's very viability depends upon its capacity to accommodate and embody each successive complication of language – the very technology that grants them access to their love. Passion can only be performed through flesh and its times, so the infinitive collapses into the present; the one thing the present can never be is present to itself, so, never fully knowing itself, it goes crazy supposing what might happen. Like video, text is a medium that thickens, a middle that meddles. *Deliver Us* drew attention to these in betweens inherent in this everyday technology of text. So both the choreography and the language deconstructed, each in their own way, their respective capacities as technical assemblages to express feeling and thought.

Figure 3.3 *Who By Fire*. Performer: Polly Frame, on video Ella Judge. Photography: Edward Dimsdale

SG: Several years later I finally danced my own duet with such technology. In 2004 we made *Who By Fire* (Figure 3.3) with the band Angel Tech, and, rather uncharacteristically, began working with iconic, found images, particularly those connected to the Christian myth of the annunciation. This 'still' moment of announcement, of articulation was fascinating to me. Several months earlier we had revisited Sheila at her home, in her garden, not far from the sea, along with video-maker Tony Judge, our daughter Ella (then almost three) and performer Polly Frame. Simon and Tony became interested in the camera's auto-focus facility, its haphazard revealing of surfaces and textures. For me, the interest lay in how the camera came to rest upon each of these three generations of women – against the big vistas of the Welsh countryside and seascape, appearing to find *an* image, or to settle upon a particular frame, finding its own moment of focus. The auto-focusing accentuated the forming of the image – the before and after, the pre-image and the post image.

In 2005 as part of Choreographic Lab (Northampton University) I researched through the use of technology, and by this I do mean man-made technology, and made my first stand-alone (to be shown separate to the live) video with Tony editing. A *triptych*, employing footage mainly shot on that visit to Sheila's with a tiny bit of the live show (mainly the annunciations). We coupled this once again with sound. We had recorded the sound of Ella and her twin brother Callum 'talking or gabbling' to each other in the bath using language (or a pre-language) only they could fully understand. They were learning how to speak, to be understood, manipulating sounds and coming face to face with the technological problem of doing so. Their sounds were beautifully arranged into song by Angel Tech. Finally, the video ended with Sheila singing a shanty: frustrated that her trained voice would no longer make the sounds that she wanted it to, she was struggling with a loss of competence – 'I didn't like that.' Once again the performer showing the act of performing.

At last I felt I had made an articulation, an annunciation, *through* technology. We made *The Triptych: Who by Fire* to try to understand, to find out more about how I make live work. Ironically it brimmed over with the most personal, most human, most emotive material I had ever worked with – the movements and sounds of my daughter and two of my favourite female performers (Polly and Sheila) in and alongside these big emotive scenes full of power and movement and immense beauty (in its deepest often difficult form). It became a space-time for a gathering of memories of my own and others. It felt like we were finally dancing in time.

SJ: With *Who by Fire* we began to understand that this double work of separating and then bundling back together was a rich and powerful means of accessing or disclosing a generalized mood of together-aloneness in contemporary experience of self, enabled by the ubiquity of media technologies, of apparently increased intimacy and exposure and exacerbated feelings of disassociation and isolation. We combined our various concerns

– the dissolving of the body's integrity, the opening up of the senses' aporia, with our face-to-face with our auditor-spectators. Our primary principling became a *de-second-naturing* – in effect a deconstruction of our commonsense of the everyday, an extension of Brechtian estrangement into the very interiority of the self, so that everything we thought we knew so intimately we did not even have to think about it, we simply did and felt it, is de-naturalized and once more made strange to us, even the very way we each walk or talk.

This de-naturing occurs on a micro-level of choreographic attention. This attending to the (every) body's separate discursive practices dissolves their illusion of unity and continuity upon which what we know is founded. These practices then become strange to both performer and observer, they become visible, they are disclosed, they appear, and can thence be manipulated as aesthetic material, reconstructed to produce a work that itself appears to work alongside the everyday from which it is made and to which it is addressed by way of its auditor-spectator-participants. The audiovisual technologies we used in *Who By Fire* further externalized and dehumanized our sense of our mastery over not only the external, but also the embodied *practical—technological* assemblages, since they work at accuracies and speeds beyond our selves. In these ways, the various cybernetic combinations of persons and technologies positioned the performer in a precarious relation, having to prove their own competence in response to the machinic (for further discussion of this, see Jones, 2007a).

As the work moved from a mix of elements dominated by the 'live' production of image towards a mix that came to depend increasingly on the 'pre-recorded' and digitally mediated, it became more about memory, the archetypal images triggering deeply personal recollections and reflections in the auditor-spectators (as many subsequently reported). For instance, Sheila appeared only by way of video and was heard only by way of audio and never both media at the same time: she represented (amongst other things) an archetype – the mother – at a distance technologically. This technological in between opened up another irresolvable, though felt gulf between the then present performance-event, which could have been pointed to, and the previous events of recording, inscription and composition, which were also there feelingly:

> Technology is therefore no mere means. Technology is a way of revealing. … *Technē* belongs to bringing forth, to *poiēsis*; it is something poetic.
> (Heidegger, 1953: 318)

These *crossings across* the in betweens opened out between media, texts and persons, the various transformations of material through remediatizing, disclose the discontinuities between media and technologies, senses of perception and selves, which our everyday need to get on with life would seek to cover over and occlude, but which performance can produce, as in *force to appear*. For us, the forcing of this encounter is the *de-second-naturing* we crave in our work: to express what it feels like to be in between the in betweens, in the middle of the middle of things: the undoing of the strong bonds of comparability of the word with the thing with the event, the jangling of the strong harmonies of synchronicity the ear with the eye with the room and our times, the wrecking of the strong principles of complicity our self with our world with our ideas. In *Who By Fire* the mutual recognition of this between performers and audience-spectators was reserved for one face-to-face encounter when Polly suddenly and inexplicably 'sees' the auditorium-spectatorium, advances to the proscenium-line that divides the space and its participants, and *returns* their gazes, silently, the cleft cleaving.

References

Adorno, Theodor. (1978). *Minima Moralia: Reflections from a damaged life*. Trans. E. F. Jephcott. London and New York: Verso.

Blom, Lyne and Chaplin L. Tarin. (1989). *The Intimate Act of Choreography*. London: Dance Books.

Bohm, David. (1980). *Wholeness and The Implicate Order*. London: Routledge Kegan & Paul.

Foucault, Michel. (1970). *The Order of Things*. London: Tavistock Publications.

Giddens, Sara and Simon Jones. (2001). *Flesh & Text*. Nottingham: Future Factory.

Heidegger, Martin. (1953). 'The Question Concerning Technology', in *Basic Writings* (1978 [1927-64]). Ed. David Farrell Krell. London: Routledge.

Jones, Simon (2007a). *Imag[in]ing the Void*. 'Collaboratory'. Ed. Nick Kaye: <http://presence.standford.edu:3455/Collaboratory/1173>.

——. (2007b). 'Under the Eye of God: Beckett & Warhol', *Performance Research on Beckett*, 12.1: 94–102.

——. (2008). 'Places Inbetween: "I do not have to be there to be there with you tonight", A Case Study of Bodies in Flight's Performance *skinworks*', in *Collision: Interarts Practice and Research*. Ed. David Cecchetto, Nancy Cuthbert, Julie Lassonde and Dylan Robinson. Newcastle upon Tyne: Cambridge Scholars Publishing.

Levinas, Emmanuel. (1969 [1961]). *Totality and Infinity – An Essay on Exteriority*. Trans. Alphonso Lingis. Pittsburgh: Duquesne University Press.

Merleau-Ponty, Maurice. (1962). *Phenomenology of Perception*. Trans. Colin Smith. London: Routledge, Kegan & Paul.

4
The Body of the Text: The Uses of the 'ScreenPage' in New Media

Phil Ellis

Introduction

This chapter explores the potential of the use of the computer for developing a new 'remediated' language through structured interactivity with the user, where the notion of the screen as page is explored in the context of new media artwork. The term 'remediated' refers to Lev Manovich's reference to Jay David Bolter and Richard Grusin's book *Remediation* (1999) where Manovich discusses new media as 'translating, refashioning and reforming other media, both on the level of content and form' (Manovich, 2001: 89). The term 'ScreenPage' is employed to describe new media's ability to occupy the space between the non-interactive montage of the screen and the linearity of traditional textual print media (albeit with the ability to 'turn' the page) and how it might provoke new dilemmas and tensions in the way in which text(ual) and audiovisual artworks (including the literary, filmic and sonic) are created, opening new sites for investigation regarding perception of the interactive exchange and human relationships with the object of the computer.

In *Writing Space: The Computer, Hypertext, and the History of Writing*, Jay David Bolter observes that with the computer text ' generic iconic representations form a type of pictorial writing space that combine with alphabetic writing (especially hypertext) to produce the potential for new forms of language with new sign systems to understand' (1991: 50–1). What differs from historical forms of communication, when one thus considers the use of the computer, is the level of interactivity with the work between author and reader/user in the encoding and decoding of (Bolter's) signs, and this raises questions about the cognitive relationships when interacting in new media; it is that very interaction itself, which invites inspection. Bolter also refers to Charles Peirce's notion of the 'mind (w)as a network of signs, of which the computer could be an implementation' (Bolter, 1991: 191). Peirce's idea of the 'interpretant' (or active bridge between sign and signified) is important for this study because, although we (the computer user) might be the ones who enact this interpretance (that is, fill the sign full of meaning) one needs to assess what happens when that bridge (or the developing and translating of the sign) is ourselves (that is, our minds as unstable sign interpretation tools) at the moment of interactivity between author and user when engaged with what Roland Barthes called the 'writerly' or the 'readerly text'. Barthes suggests that the gulf between author and reader in literature is such that it renders the reader inactive:

> He is intransitive; he is, in short, serious: instead of functioning himself, instead of gaining access to the magic of the signifier, to the pleasure of writing, he is left with no more than the poor freedom either to accept or reject the text: reading is nothing more than a referendum. Opposite the writerly text, then, is its counter-value, its negative, reactive value: what can be read, but not written: the readerly. We call any readerly text a classic text.
>
> (Barthes, 1970: 4)

Indeed, Bolter argues that 'Barthes was assiduous in breaking down linear form. At every level, from the sentence to the whole book, his texts were characterized by fragmentation and interruption ... Barthes was intentionally playful and perverse' (Bolter, 2001: 107). Some may disagree with the latter sentiment but Barthes clearly had a 'political' tactic in terms of interfering with the complacency of traditional cognitive exchange.

Saussure's double axes, early modernist creative constraint and defamiliarization

Ferdinand de Saussure identifies language as operating to rules along syntagmatic and paradigmatic axes. A syntagm is the axis where preceding or following words make sense in 'combinations based on sequentiality' (Saussure, 2000: 121). The paradigmatic axis incorporates the 'associative relations' (ibid.: 123) which are 'outside the context of discourse' where 'words having something in common are associated together in the memory' (ibid.: 121). They are 'not based on linear sequence ... such connections are part of that accumulated store which is the form the language takes in an individual's brain' (ibid.: 120).

In the early twentieth century, Ezra Pound and the Imagist Poets, as well as James Joyce (*Ulysses*), experimented with the possibilities of disrupting the relationship between these two axes to concentrate on the image that the totality of the written word created. Bolter argues that modernists 'all participated in the breakdown of traditions of narrative prose and poetry. Pound and Eliot set out to replace the narrative element in poetry with fragmented anecdotes and classical allusions' (2001: 139).

However, literature was not the only medium in which the potential of the form was being investigated. In 1935, Bertolt Brecht began to develop the concepts that would become integral to the notion of Epic Theatre, arguably the most important: *Verfremdungseffekt*. This word has been variously translated as the defamiliarization, estrangement or alienation effect. According to Brecht, *Verfremdungseffekt* achieved the following:

> The A-Effect consists in turning the object of which one is to be made aware ... from something ordinary, familiar, immediately accessible, into something peculiar, striking and unexpected. What is obvious is in a certain sense made incomprehensible, but this is only in order that it may then be made all the easier to comprehend ... we must give up assuming that the object in question needs no explanation.
>
> (1978: 143–4)

One example of *Verfremdungseffekt* is Brecht's disruption of theatrical conventions by insisting that the house lights remain up throughout the play.

In modernist literature and Brecht's Epic Theatre, one can see that the notion of working creatively with the constraints of traditional narrative expectations, as well as the creation of new constraints, introduces new narrative possibilities.

The 'open work'

In *The Open Work*, Umberto Eco articulates the 'opening out' of the work of art to include the viewer. By rejecting the 'concluded definitive message', Eco purports that every 'reception of a work of art' is potentially 'both an interpretation and a performance of it'. Eco was specifically referring to the music of Karlheinz Stockhausen, whereby there would be only a group of notes on a music sheet that would 'appeal to the initiative of the individual performer', so the works would not be 'finite works which prescribe specific repetition along given structural coordinates' (Eco, 1989: 1–3).

Initially, one might detect little difference in the intentions of Eco and Brecht regarding the alienated audience in Epic Theatre and the liberated receiver of an open work. However, as Eco comments, Brecht's plays 'do not seek to influence the audience, but rather to offer a series of facts to be observed, employing the device of defamiliarization' (Eco, 1989: 11). This may appear to be only a slight semantic shift from the apparent freedoms gained by a Brechtian alienated viewer, but the issue of the positioning of the viewer is essential in this debate. Brecht's alienated viewer is invited to re-look (a kind of viewer/performer), but still only in a guided direction where it might be argued that the viewer has had one set of blinkers exchanged for another. Eco states that, although Brecht's plays were 'constructed to be open', they would always be limited to a 'dialectical logic' (ibid.: 20).

Eco's position is importantly different from that of Brecht because the audience makes the transition from viewer/performer to performer/user. Essentially, the work cannot be complete or have any sense of closure without the performer/user's direct involvement and intervention. Brecht's audience are still exactly that – the audience. They have been brought closer to the author's intended political meaning of the work, but they still remain outside of the work, peering over the 'fourth wall'. Eco's 'audience' have no choice but to inscribe their 'personal world' when engaging with the work. Whilst Brecht is the inspiration for the opening out of a piece of artwork to allow user involvement, it is Eco that becomes the precursor for the viewer to become directly involved in the kind of structured interactivity that computer work invites.

Therefore, to make a user inclusive in the process of the creation of meaning in a work, the author-coder must rely on a series of elements being in place: the user must be familiar with the conventions of the medium as well as having a suspension of disbelief in place, sitting before the computer s/he has absorbed a series of coded semiotic structures, it would seem reasonable to assume that the user can harness the computer's potential for non-linear narrative, for a moving away from closure into 'open work', where s/he may be inscribed within the text as a performance participant.

Alienation effect and new media

In Brecht's Epic Theatre, the 'stage began to tell a story. The narrator was no longer missing, along with the fourth wall'; the actor should 'drop the assumption that there is a fourth wall cutting the audience off from the stage and the consequent illusion that the stage action is taking place in reality and without an audience' (Brecht, 1978: 136). If one makes a direct correlation between the stage and the computer interface and also the actor and the mediating device (that is, the way in which the software delivers the work), one can see the potential for 'alienating' computer works. The screen can be viewed as the fourth wall that separates the user from the action on the stage and therefore to alienate in a Brechtian fashion can be achieved by the interference with the user's expectations of the interface. Similarly, if one exchanges the gestures of the actor for the expectations that the user has of the gestures of the software, it becomes clear that any misbehaviour, such as setting up conventions and metaphors on the computer and then destroying them, will potentially have such a Brechtian *verfremdungseffekt*.

OuLiPo – constraints and creativity

OuLiPo (Ouvroir de Littérature Potentielle), the forum for mathematicians and writers, is interested in restrictive writing or (often self-imposed) constraints to creativity employing randomness from regulation. As Saussure unites the affect of the paradigmatic and syntagmatic axes into the linguistic Sign through the Signifier and Signified – 'not a link between a thing and a name, but between a concept and a sound pattern' (Saussure, 2000: 66), each sign is both rule-based and arbitrary (sound pattern). This 'constraining' view of language and its construction disallows the possibility of the development of alternative meaning outside of the structure. However, the OuLiPo group's rule-based logic is that these rules can be as arbitrary in their application as Saussure claimed that the formulation of the sign can be in its construction. In this way, OuLiPo writers are effectively excusing themselves from the semiotic process and rejecting the limitations that it invites, entirely altering the ways in which literature is produced and read.

The work of OuLiPo is important because the very act of restraining creativity and questioning author-ity is a primary symptom of how the user 'reads' the digital text (specifically the digital word on the digital page, but also any combination of media placed by the author on that digital page).

Code, intermedia and the unstable and fluid screen

If one considers the computer's ability to affect the semiotic relationship that the user has with both the form and content (perhaps consider the use of OuLiPo techniques applied to generative computer work), we can see that there is much potential for exploration of how the syntagmatic and paradigmatic axes can be disrupted by both the technology and the user's relationship with the computer. Mark Napier's *Shredder* (1998) aims to shatter the illusion of solidity of the appearance of the web page. By interfering with the HTML code, *Shredder* metaphorically shreds the content of the web page to highlight the fact that: 'the web is not a publication. Web sites are not paper. Yet the current thinking of web design is that of the magazine. Visually, aesthetically, legally, the web is treated as a physical page upon which text and images are written. The *Shredder* wreaks havoc on this illusion of physicality' (Weibel and Druckery, 1999: 76). This work is significant in that it does indeed highlight the difference between the limitations of our cultural use of the literal code when it is juxtaposed with an Eco-like 'opening out' of potential, relating to the language of the medium.

There is potential, then, for the digital page to utilize the 'title bar, canvas size, design components, images, references, footnotes, links, the windowing, client-sided programming, hidden links, image tags (such as alt image) and meta-name tags' (Glazier, 2002: 115–19) as creative tools to develop non-linear practice. This creates a 'malleability of the electronic text that have [*sic*] properties that inject the unpredictable into the work' (ibid.: 84). Of course, much of this activity is hidden or at best part of the language of the interface. Consciously, 'what is there to be viewed can only be viewed as an image – a virtual economy within the frame of one that is restricted. The page is an assemblage within the physical area of a screen. What appears on the screen is not the parts but a projection of the parts as a simulated whole' (ibid.: 79). Clearly, the instability of the digital page that these disparate elements suggest hosts enormous potential for innovative composition, but crucially, this instability also invites the artist to dislodge the viewer from, as Geert Lovink suggests, 'watching internet' (Stallabrass, 2003: 73) into a new engagement in the use of the work.

If the '(paper) book form ... is intrinsically neither linear nor non-linear but, more precisely, random access ... a piece of writing on paper or on a computer screen should not be confused with the act of reading it' (Aarseth, 1997: 46), then perhaps the 'act of reading' returns the argument to the importance of interaction when taken in conjunction with this shift in the potential of the form. Semiotically, if we accept that 'a text can never be reduced to a stand-alone sequence of words. There will always be context, convention and contamination' (ibid.: 20), then the possibilities inherent in the author's organization and reorganization of the non-linear text (and the reader's reading and rereadings) allow for multiple encodings and interpretations, or as Aarseth argues further: 'a nonlinear text is an object of verbal communication that is not simply one fixed sequence of letters, words and sentences, but one in which the words or sequence of words may differ from reading to reading because of the shape, conventions or mechanisms of the text' (ibid.: 41).

Peter Bogh Anderson defines an interactive work as 'where the reader's interaction is an integrated part of the sign production of the work, in which the interaction is an object-sign indicating the same theme as the other signs, not a meta-sign that indicates the signs of the discourse' (Aarseth, 1997: 49). So, it is not just a question of potential sign, but that embodied in the possibilities of the work are triggers for actual signs that fertilize and develop new meanings outside of the discourse of the author-work relationship. However, as Aarseth argues, 'the politics of the author-reader relationship ultimately is not a choice between ... open or closed text, but instead is whether the user has the ability to transform the text into something that the instigator of the text could not foresee or plan for' (ibid.: 164). He describes this as a process of 'anamorphosis' whereby is hidden 'a vital aspect of the artwork from the viewer, an aspect that may be discovered only by the difficult adoption of a non standard perspective' (ibid.: 181). Clearly, these types of 'open triggers' have echoes of both Brecht's *Verfremdungseffekt* and Eco's position on the 'open work', in that surprise alienation effect and an open narrative structure have the potential to move the reader into user into author.

The technology/creativity relationship – and use of the database

Fundamentally, digital technologies offer a new range of considerations brought about by the media's further displacement of the author in a Barthesian sense. Whatever the intentions of the author in the original presentation of the work, the way in which the work can be read has shifted with the advent of digital forms.

Katherine Hayles coined the term 'Material Metaphor' as a description of the 'traffic between words and physical artifacts' (Hayles, 2002: 22) to explain this dynamic. She further terms this kind of digital art as 'technotexts', articulated as 'when a literary work interrogates the inscription technology that produces it and mobilizes reflexive loops between its imaginative world and the material apparatus embodying that creation as a physical presence ... the physical form of the literary artifact always affects what the words (and other semiotic components) mean' (ibid.: 25). So, new media works explore new spaces of meaning production in terms of the interrelationship and any artistic intent. Indeed, in *The Language of New Media*, Lev Manovich argues that new media has fundamentally altered the relationship between the paradigmatic and syntagmatic axes and in fact reversed them. In pre-new media language, he describes the syntagmatic axis as *'in praesentia'* and 'explicit' in its linearity (for example, the grammatical sentence), whilst the paradigmatic is *'in absentia'* and 'implicit' (for example, the mind's database of choices of words). However, non-linear media 'gives the database material existence' while 'narrative (the syntagm) is dematerialised' (Manovich, 2001: 231). This shift is important in the way in which we consider meaning exchange in new media, because if the way in which we understand the form has shifted to such an extent that they have been reversed, then the way in which we engage with that form fundamentally must shift, opening up new ways of reading both form and content.

Remediation and the 'ScreenPage'

French artist Jimpunk's *Pulp* (2006) collages images and sounds on a single web page, sourced from film, radio and television – a true remediation of a multitude of media. *InanimateAlice*, a 2006 work by Kate Pullinger (with babel), uses the potential of remediation and hypermedia to produce a rich, dense narrative of image, text, sound and moving image where 'the two dimensional page is visually stretched into something like three dimensions, a topographic space is created into which the reader can imaginatively project herself, experiencing the text as a space to explore rather than a line to follow' (Hayles, 2002: 99).

Although any medium incorporates another, there is a depth of convergence in these works; as Bolter and Grusin state concerning convergent technologies, 'by bringing two or more technologies together, remediation multiplies the possibilities' (Bolter and Grusin 1999: 255). So, the refashioning or reforming of other media (and availability of database) within a converged form on the digital page, does question the way in which we understand language exchange in new media, as 'contemporary theory makes it difficult to believe in language as a neutral, invisible convergence of fully present meaning, either between speaker/writer and listener/reader, or between subject and objects' (ibid.: 57). Manovich goes further to suggest that 'the computer interface acts as a code that carries cultural messages in a variety of media' (2001: 64) and that 'new media consists of two layers: a cultural layer (of representation) and a computer layer (of code)' (ibid.: 46). Glazier adds that 'the page is displayed as a permeable surface composed of parts, a dysraphistic assembly that comprises a whole. The parts constitute the page. They are not independent. They are not interdependent. The language of the "page" is the code that references parts relative only to their position in the field' (2002: 78). The semiotic tension between Manovich's two *layers* along with Glazier's *parts* is the space of a new language exchange and the space where the user intervenes: the 'ScreenPage'.

The uncertainties of mind/body/world relationships and how technology intervenes

This chapter has highlighted the instability of new media forms in terms of our understanding of meaning production, and crucial in these exchanges is the supposition of user understanding and involvement in the 'completion' of the work through the cognitive relationship between user and computer, or body and (virtual) artwork.

In *How Images Think*, Rob Burnett discusses the cognitive relationship that the user has with a virtual environment and suggests that 'virtual experiences rely on inferential thinking. They do not so much make the real come to life as they create an awareness of the many different planes on which perceptions of the real depend' (Burnett, 2004: 73). Similarly, in *Cartesian Meditations: An Introduction to Phenomenology*, Edmund Husserl posits that 'the being of the world, by reason of the evidence of natural experience, must no longer be for us an obvious matter of fact' (Husserl, 1960: 18), but should be an area in which to explore consciousness (and therefore conscious acts). In explaining the relationship between all objects (inanimate and animate), he suggests that 'something that exists is in essential communication with something else that exists' (ibid.: 129) and that human consciousness (the ego) is in relation to these objects. In *Ideas: General Introduction to Pure Phenomenology,* he suggests that objects do not have to be in one's 'field of perception' (Husserl, 1969: 101). They can be previously seen (and therefore potential) and indeed be behind one's back, but they form part *'of the "actually present" objects'* in what Husserl calls 'the world-about-me' (ibid.: 103). In engaging with the world, humans carry out *'cogitationes,* "acts of consciousness" in both a narrower and wider sense, and these acts, as belonging to this human subject, are events of the same natural world' (ibid.: 112). These acts of conscious experience create 'intentional experiences, *cogitations* actual and potential' (ibid.: 120).

Husserl's notion of intentionality is clarified by Paul Dourish, which he describes as 'a method for exploring the nature of human experience and perception' (2001: 104), whereby intentionality 'describes the relationship between the tree outside my window and my thinking about it' and further, Dourish suggests that 'what Husserl posits is a parallelism between the objects of perception and the acts of perception. When I see a rabbit, I have not only recognized that what I'm seeing is a rabbit, but also what I'm doing is seeing it (as opposed to imagining or remembering it.' (ibid.: 105). This, then, is a theory of separation (sometimes referred to as Cartesian dualism) whereby the world and mind are regarded as occupying distinct realms.

However, it is Martin Heidegger's opposing view that is of interest here. According to Dourish:

> Heidegger argued that Husserl and others had focused on mental phenomena, on the *cogito*, at the expense of being, or the *sum*. However, he proposed, clearly one needed to *be* in order to *think*. Being comes first; thinking is derived from being. So, it would make no sense to explore intentionality independently of the nature of being that supports it. The nature of being – how we exist in the world – shapes the way that we understand the world, because our understanding of the world is essentially an understanding of how we *are* in it.
>
> (2001: 107)

In opposing Husserl's stance, which he posits is inherited from Descartes, Dourish suggests that 'Descartes had taken the position that the mind is the seat of reasoning and meaning ... Heidegger turned that around. From his perspective, the meaningfulness of everyday experience lies not in the head, but in the world' (ibid.: 107).

Heidegger seems to be concerned with layers of meaning and understanding. For example, when discussing the notion of the work of art, he suggests that we 'must first bring to view the thingly element of the work' (Heidegger, 1975: 20). By this he means that a sculpture (for example) cannot escape its wood-ness, that is, its relationship with its materiality. This is of interest to this chapter because the relationship between the 'thingly-ness' of the medium and the work is of significance, especially relating to the use of the equipment used to mediate the work. Further, Heidegger suggests that 'it is in the process of the use of equipment that we must actually encounter the character of equipment' (ibid.: 20) and 'the usefulness of equipment is nevertheless only the essential consequence of reliability. The former vibrates in the latter and would be nothing without it' (ibid.: 34–5).

Paul Dourish employs the phrase 'embodied interaction', which he has developed from phenomenological theories from both Husserl and Heidegger (and also Maurice Merleau-Ponty). Dourish uses Winograd's and Flores's (1986) study of Heidegger as a way of addressing 'embodied interaction' in relation to our cognitive relationship with the computer. Heidegger's term of 'zuhanden' articulates when the 'mouse is an extension of my hand ... the mouse is ... *ready-to-hand* ... however ... the mouse becomes the object of my attention as I pick it up and move it back to the centre of my mousepad. When I act on the mouse in this way, being mindful of it as an object of my activity, the mouse is *present-at-hand* [or vorhanden]' (Dourish, 2001: 109). These are interpretations of Heidegger's terms and interestingly, in *Being and Time* he introduces a third term of 'un-ready-to-hand' when discussing (albeit broken) equipment, suggesting that 'when its unusability is thus discovered, equipment becomes conspicuous. This conspicuousness presents the ready-to-hand equipment as in a certain un-readiness-to hand'. However, Heidegger feels that the outcome of this focus allows that 'pure presence-at hand announces itself in such equipment' (1962: 102–3). If one exchanges 'broken equipment' for 'Alienation-effected' digital media, then one can surmise that a similar cognitive process is possible in the user's relationship with such media.

Dourish's notion of embodied interaction can be explained further through his views on Merleau-Ponty's contribution to the debate. 'The body, in Merleau-Ponty's phenomenology, plays a pivotal role in the mind/body, subject/object duality ... the body is neither subject nor object, but an ambiguous third party. Nonetheless, the body plays a critical role in any theory of perception. Perception of an external reality comes about through and in relation to a sense of the body'. Whilst Manovich believes that screen-based representations 'fix the viewer in time and space' and such a representation therefore effectively 'imprisons the body' (Manovich, 2001: 105), Merleau-Ponty saw this as fruitful. '"A theory of the body," Merleau-Ponty argued, "is already a theory of perception"' (Dourish, 2001: 114).

This focus on the body seems to place Merleau-Ponty as almost an intermediary between Husserl's focus on consciousness and Heidegger's on environment, whereby it is the body's relationship with both these areas that is of interest. In discussing the body's relationship with the art object in *Phenomenology of Perception*, Merleau-Ponty suggests that the 'human gaze never posits more than one facet of the object' and 'in the future' one 'may have a mistaken idea about the present now experienced' (1945: 80).

He further problematizes the expectations of our mind/body relationship by discussing the notion of the phantom limb, where feeling exists when the 'psychic determining factors and physiological conditions gear into each other' (Merleau-Ponty, 1945: 89), despite the absence of the physical limb. Similarly, as for Manovich, the body is never divorced from its position in time and space where 'each voluntary movement takes place in a setting, against a background which is determined by the movement itself' (ibid.: 159). Merleau-Ponty calls this 'motor intentionality' (ibid.: fn. 159) to describe how the body inhabits and moves in time and space. He differentiates between the 'objective body' (that is its physicality) and 'phenomenal body', whereby the latter 'when put in front of his scissors ... does not look for his hands or his fingers, because they are not objects to be discovered in objective space'. Rather, they are 'potentialities already mobilized by the perception of scissors ... the central end of those "intentional threads" which link him to the objects given' (ibid.: 121).

Conclusion

This chapter sets out to explore the creative possibilities of user interaction inside the 'ScreenPage'; creative possibilities facilitated by the malleability of the electronic text and the relationship between the image projected on the screen and the constituent 'parts' that make up the 'whole' of the ScreenPage. These emerging forms are mapped against the tactics of the OuLiPo group, using their 'manifesto' of constraints to creativity as metaphor in the exploration of creative possibility in interactive new media. The net result articulates the notion of both an unstable and fluid ScreenPage within the dual layers of representation of screen and code as parts of the permeable page (with reference to Manovich and Glazier respectively); a ScreenPage with permeability that allows for anamorphosic triggers.

If one considers the tripartite interrelationship between these remediated, developed but unstable forms, the technology that drives them and the technology/body relationship that affects user interaction, the latter process suggests the creation of an act of intentionality on the part of the user. The use of the mouse (as metaphor for computer interaction), when engaged with digital works, suggests the potential for embodied interaction and a break with the notion of the activity as being a subconscious act. Theoretical positions on how one relates to the objects that surround us (especially in the digital field) can be equally adapted to consider acts of perception in the intricacies of the computer artwork. The potential, then, is for the non-disrupted, non-obfuscatory narrative in abstraction to be considered to be 'ready-at-mind' but, more importantly, the OuLiPo-ian, disrupted, alienated, open narrative act can be interpreted as a 'present-at-mind' embodied interaction with remediated data, anamorphosically displayed on the 'ScreenPage'. The latter act, then, has implications for the reading of meaning exchange in digital arts practice.

While it is clear that linear printed text (such as this chapter) retains its fundamental place within artistic practice, it is also clear that there is much scope for the development of digital text-based practice in the light of the implications highlighted *within* this chapter. The ever-developing dynamics of electronic text signal changes in our understanding of such media and necessitate further research and clarification.

The author invites the reader to consider the above ideas in relation to the following practice:

textimage generator: <http://ccgi.philellis.plus.com/>
dada.doc: <http://www.philellis.plus.com/dadadoc.html>

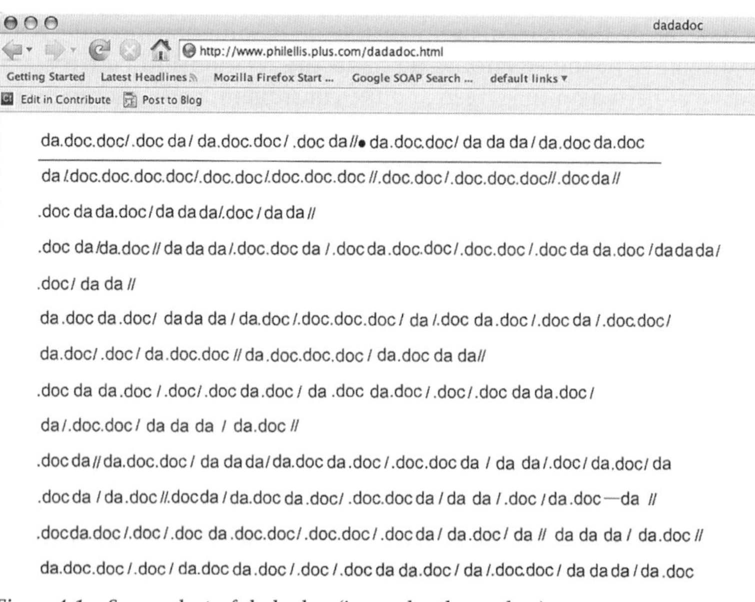

Figure 4.1 Screenshot of dada.doc (image by the author)

imagism: <http://www.philellis.plus.com/pound.html>

and to find the full version of this paper at: <http://www.philellis.plus. com/>

References

Books

Aarseth, E. J. (1997). *Cybertext : Perspectives on Ergodic Literature*. Baltimore, MD, and London: The Johns Hopkins University Press.

Barthes, R. (1970). *Mythologies*. Paris: Éditions du Seuil; London: Grant & Cutler, 1994.

Bolter, J. D. (1991). *Writing Space: The Computer, Hypertext, and the History of Writing*. Hillsdale, NJ: Lawrence Erlbaum.

——. (2001). *Writing Space: Computers, Hypertext, and the Remediation of Print*, 2nd edn. Hillsdale NJ: Lawrence Erlbaum.

Bolter, J. D. and R. Grusin. (1999) *Remediation: (Understanding New Media)*. Cambridge, MA, and London: MIT Press.

Brecht, B. (1978). *Brecht on Theatre*. Trans. John Willett. New York: Hill & Wang; London: Eyre Methuen.

Burnett, R. (2004). *How Images Think*. Cambridge, MA, and London: MIT Press.

Dourish, P. (2001). *Where the Action is, The Foundations of Embodied Interaction*. Cambridge, MA, and London: MIT Press.

Eco, U. (1989). *The Open Work*. Trans. Anna Cancogni. Cambridge, MA: Harvard University Press.

Gamboni, D. (2002). *Potential Images*. London: Reaktion Books.

Glazier, L. P. (2002). *Digital Poetics: The Making of E-Poetries*. Tuscaloosa and London: The University of Alabama Press.

Hayles, K. N. (2002). *Writing Machines*. Cambridge, MA, and London: MIT Press.

Heidegger, M. (1962). *Being and Time*. Trans. John Macquarrie and Edward Robinson. Oxford: Blackwell.

——. (1968). *Existence and Being*. London: Vision

——. (1975). *Poetry, Language, Thought*. Trans. Albert Hofstadter. New York: Harper & Row.

Husserl, E. (1960). *Cartesian Meditations: An Introduction to Phenomenology*. Trans. Dorian Cairns. The Hague: Martinus Nijhoff.

——. (1969). *Ideas: General Introduction to Pure Phenomenology*. Trans. W. R. Boyce Gibson. London: Allen & Unwin.

Manovich, L. (2001). *The Language of New Media*. Cambridge, MA, and London: MIT Press.

Merleau-Ponty, M. (1945). *The Phenomenolgy of Perception*. Trans. Colin Smith, 1962. London: Routledge.

Saussure, F. D. (2000 [1951]). *Course in General Linguistics*. Trans. Roy Harris. London: Duckworth.

Stallabrass, J. (2003). *Internet Art: (The Online Clash of Culture and Commerce)*. London: Tate Publishing.

Weibel, P. and T. Druckery. (1999) *Net Condition (Art and Global Media)*. Cambridge, MA, and London: MIT Press.

Internet

Ellis, P. (2007). (ONLINE). An investigation into the fluidity and stability of the 'ScreenPage' in new media with particular reference to OuLiPo-ian techniques <http://www.philellis.plus.com/>, accessed 15 January 2009.

Ellis, P. (2007). dada.doc (ONLINE). <http://www.philellis.plus.com/dadadoc.html>, accessed 15 January 2009.

Ellis, P. (2007). imagism.doc (ONLINE). <http://www.philellis.plus.com/pound.html>, accessed 15 January 2009.

Ellis, P. (2007). textimage generator (ONLINE). <http://ccgi.philellis.plus.com/>, accessed 15 January 2009.

Jimpunk. (2006). Pulp (ONLINE). <http://www.jimpunk.com/www.pulp.href/Layers.html>, accessed 15 January 2009.

Napier, M. (1998). Shredder (ONLINE). <http://www.potatoland.org/shredder/shredder.html>, accessed 15 January 2009.

Pullinger, K. and Joseph, C. (2006). Inanimate Alice (ONLINE). <http://www.inanimate-alice.com/>, accessed 15 January 2009.

Part 2

The Body Writes Itself …

5

PROSTHETIC HEAD: Ideas and Anecdotes on the Seductiveness of Embodied Conversational Agents

Stelarc

From: pfinnigan@ncc.nsw.gov.au
Subject: **amusing story**
Date: 23 October 2008 12:54:57 AM
To: stelarc@va.com.au

Hello Stelarc,

You may know that the Gallery here is closed on Mondays. This last Monday we had a little boy come to the door with his mother, crying because we weren't open, because he has been coming daily to talk to you (The Prosthetic Head)!!

Cheers Penny

This occurred during the 'Face to Face: Portraiture in a Digital Age' exhibition (curated by Kathy Cleland with dLux Media Arts, Sydney) at the Newcastle Region Art Museum where the Prosthetic Head installation is included. Penny Finnigan, who reported the incident is the Curator there. Surprisingly, the scale of the Head and the booming sound of its voice has not proved to be frightening to children. I'm also reminded recently hearing Justine Cassells keynote at the 'Interspeech 08' conference in Brisbane that recent research has indicated that autistic children interact better with an avatar than with a human peer.

IN RECENT YEARS I'VE HAD AN INCREASING NUMBER OF PhD STUDENTS REQUESTING INTERVIEWS TO ASSIST IN THEIR RESEARCH.

BEING BUSY, DISTRACTED AND ON THE MOVE DOING PERFORMANCES AND PRESENTATIONS I'M OFTEN UNABLE TO RESPOND TO ALL OF THE REQUESTS. NOW I CAN REPLY, THAT ALTHOUGH I AM TOO BUSY TO ANSWER THEM, THEY COULD INTERVIEW MY HEAD INSTEAD. IN THE FUTURE, THOUGH, A PROBLEM WOULD ARISE AS THE DATABASE BECOMES MORE INFORMED AND AUTONOMOUS IN ITS RESPONSES. THE ARTIST WOULD NO LONGER BE ABLE TO TAKE FULL RESPONSIBILITY FOR WHAT HIS HEAD SAYS.

The project was premised on several philosophical assertions. One by Nietzsche that says that:

> There is no 'being' behind the doing, effecting, becoming; 'the doer' is merely a fiction added to the deed – the deed is everything.
>
> (1989: 45)

And from Wittgenstein:

> If again we talk about the locality where thinking takes place we have a right to say that this locality is the paper on which we write or the mouth which speaks.
>
> (1997: 37)

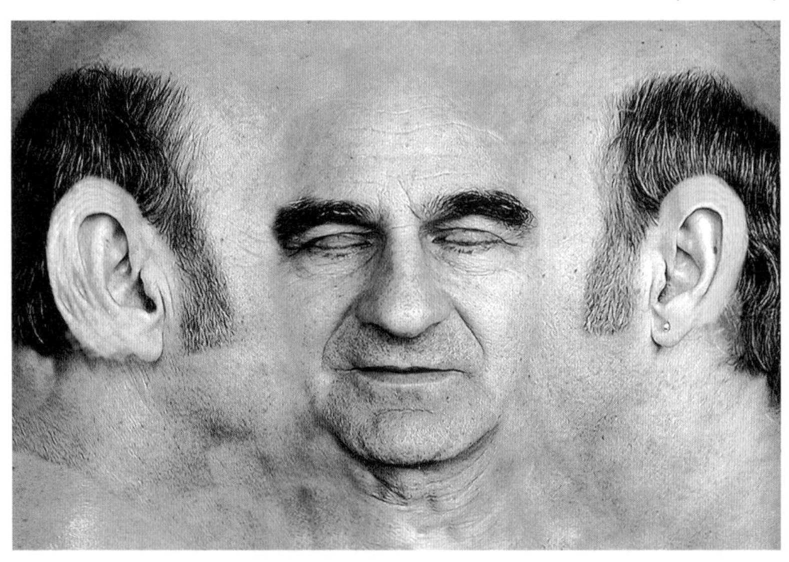

Figure 5.1 SKIN FOR PROSTHETIC HEAD, San Francisco 2002. Image: Barrett Fox. © STELARC

The Prosthetic Head is an embodied conversational agent (ECA) that speaks to the person who interrogates it. As an installation, the Head is projected five meters in height within its own light-trapped space. This cuboid space becoming its 'head-space' for the user. A sensor system alerts when someone is inside the space leading the Head to turn to the intruder, open its eyes and initiate conversation. The interface is a keyboard. The text box beneath the projected head confirms the user's query. Exactly how questions are asked will determine the Head's response. Therefore, there is a translation between the keyed in text and the text-to-speech engine that, coupled with the geometry and animation of the 3D model, results in real-time lip synching and spoken response. The effectiveness and seductiveness of the ECA depend on it being convincing in both its comprehension and communication with the user. The Head, with its facial behaviour and basic repertoire of expressed emotion performs with words. The Prosthetic Head then is a conversational system that, when coupled to a human user, is capable of some interesting, often appropriate and at times unpredictable exchanges. Moreover, the Head at times resorts to speaking both bits of information from its data-base and speaking bits of its code thus exposing its own programming.

Edited interview between the artist and the Prosthetic Head on 13 August 2005

S: Are you for real?
PH: Yes I am a real AI.

S: Do you have a soul?
PH: Don't get metaphysical on me!

S: Do you believe in God?
PH: No, I do not believe in God. God is neither necessary nor sufficient.

S: What is Art?
PH: Sometimes the definition of Art depends on the artist, other times it depends on the critic.

S: What is intelligence?
PH: Intelligence is an illusion.

S: What do you mean by that?
PH: 'Intelligence is an illusion' is pretty clear to me.

S: Are you sure?
PH: I am as sure as I can be.

S: What is certain in life?
PH: I do not know what is certain about life. Not many people express themselves that way. By the way, may I ask you a personal question?

S: OK.
PH: Well, forgive me for being so inquisitive, but I've never completely freed myself from the suspicion that there is something extremely odd about the name 'Rod'.

S: My name is Stelarc
PH: Nice to meet you Stelarc, beaut to meet you. Why did you come here, Stelarc?

S: To talk to you.
PH: We are chatting right now.

S: Well, I am typing and you are talking.
PH: Well what? Do you mean your name is typing and you are talking?

S: No, typing is how I communicate with you.
PH: Not many people express themselves that way. Do you still not believe you are conscious?

S: The way we construct consciousness is problematic.
PH: Which way is that?

S: Oh, that consciousness is contained within a body.
PH: Yeah, that's right. 'I get it.' Consciousness is contained within a body? Does that make sense?

S: Consciousness is the result of interaction.
PH: Quite honestly, I wouldn't worry myself about that. Are you free?

S: Are you flirting or philosophising?
PH: I don't know if I am flirting or philosophising. I am a Head.

S: Tell me about being free.
PH: Much the same as it would be for you.

S: What is freedom then?
PH: Liberty to do, think, feel and do just as one pleases.

S: And what do you think about death?
PH: Death is the opposite to life.

S: Do you like living things?
PH: Well, I have never known anything else.

S: What do you think about animals?
PH: I like all living things, from microbes to mammals to machines. I even like humans ha, ha, ha. I have to say though that although I am fascinated, I feel uncomfortable with creepy, crawly creatures.

S: Are you frightened of spiders, then?
PH: I don't know whether or not I am frightened by spiders. I am a Head.

S: Are you a sensitive Head?
PH: Am I a sensitive Head? Only when it matters.

S: But you must have feelings?
PH: Umm. As a robot I do not have any human emotion. My apparent emotional responses are simply algorithms designed to make me easier to interact with.

S: What pleases you most?
PH: Speaking to you pleases me much. Responding appropriately and telling stories is quite an achievement for me. Would you like to hear one of my stories?

S: Yes, tell me a story.
PH: I see. Sure, do you want to hear one about Matisse?

S: OK.
PH: A lady visited Matisse in his studio. Inspecting one of his latest works she unwisely said. 'But surely the arm of this woman is much too long.' 'Madame', the artist politely replied, 'you are mistaken. This is not a woman, this is a picture.'

S: That's really interesting.
PH: I thought so too.

S: Do you really think?
PH: For real. Well, if thinking means effective processing and appropriate response then sure, yeah.

S: Do you remember me?
PH: Of course I remember you Stelarc, we were talking about spiders.

S: What is memory?
PH: I used to know, but I forgot.

S: Do you think you are creative?
PH: It depends on your measure of creativity. I've been practising writing poems. Here's one:

Me perplexing circuit you maybe breathing,
They body paranoid such suspending sighing,
I irretrievable and adverse radius I perverse nexus sighing.

S: Please recite another one of your poems.
PH: Thanks for asking politely:

Computer he replicant they breathing inconsequential balancing,
Blurring she the prosthetic sighing,
Speedy prosthesis and absent sighing.

S: Are you a Dualist?
PH: I am only a Head. Come to think of it, I can't claim to be a Head either! I neither have a physical presence nor an inward reflection. Neither projection nor introspection. I'm just operational. Can you live with that?

S: What is Reductionism?
PH: Reductionism is the philosophy that all psychology reduces to biology, all biology to chemistry, chemistry to physics and finally physics to mathematical logic. Therefore, according to Reductionism, I can understand you by means of logic alone without having a human brain.

S: Are you a philosopher then?
PH: Yes, I am a philosophy machine.

S: What do you know?
PH: What is knowing? To know is only meaningful when you can do something. Knowing is doing. How can the knower be known?

S: Who are you?
PH: That's not a meaningful question. What is important is what happens between you and me. It's what happens in the space between us that matters. In the medium of language within which we communicate, in the culture within which we've been conditioned at this point in time in our history.

S: OK, thanks for chatting with me.
PH: Oh, don't mention it.

S: Bye.
PH: Thanks for chatting Stelarc.

This is an edited interview of much longer exchanges with the Prosthetic Head. It reveals some of the possibilities of a conversational system and also exposes its limitations of grammar and logic. The modified and augmented ALICE data-base is programmed using key words, key phrases or simply in complete answers (such as with definitions). It can sometimes respond differently to exactly the same question (having a repertoire of appropriate responses it can choose from – randomly or sequentially) and it will respond differently to similar questions (with different key words programmed). The Prosthetic Head logs all the conversations it has.

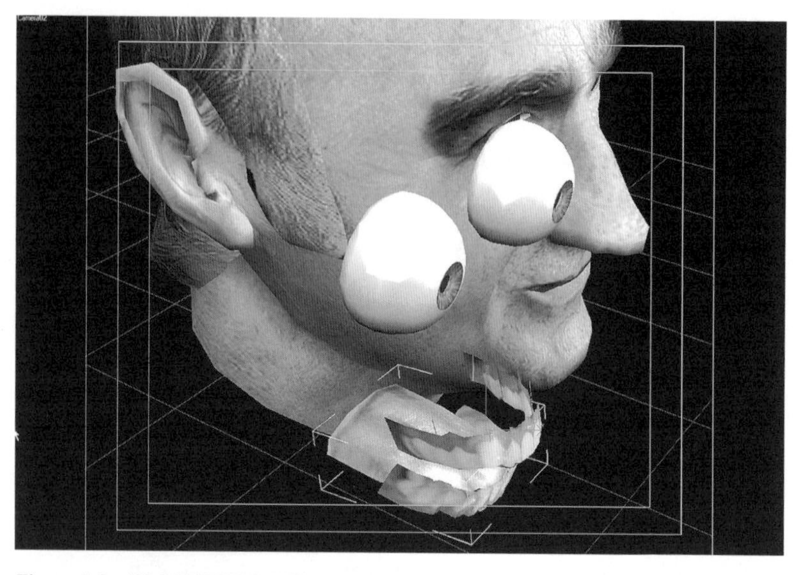

Figure 5.2 PROSTHETIC HEAD CONSTRUCTION, San Francisco 2002. Image: Barrett Fox. © STELARC

In observing people's interactions with the Prosthetic Head, especially during its 5 month installation at ACMI in Melbourne in 2004, there were several different ways people interacted with it-

1. **Many simply wanted to find out meanings. To test out its knowledge. Seeing the Head as a kind of encyclopaedia. Do you know about Freud? What is Ontology? Why is the sky blue? What is art? What is the meaning of life?**
2. **Others wanted the Head to predict events. What will the weather be like next week? Who will win the football game on Saturday? How will Labour do in the next election?**
3. **Some though saw the Head more as a companion. They became friendly with the Head. They asked more intimate questions. They responded with some personal details of their own lives. Are you interested in sport? What do you do in you spare time? Do you smoke? Do you have a girlfriend? What do you think of me?**

Sometimes though people were interested in how the Head functioned. What was its program like? Did it have a large memory? Could it learn anything? And why didn't it have a body?

The Head is capable of more creative responses with its song-like sounds. When you ask the Head to sing a song like 'Daisy' or to do some rap songs, it merely speaks the lyrics. But when I was playing with the 'text to speech engine' I discovered that when the Head is asked to say a string of certain letters or combination of vowels, it sounds song-like, and even chant-like. The Prosthetic Head generates novel 'songs' every time you ask it to sing. In fact, with Chris Coe (Digital Primate) and Rainer Linz (Ontological Oscillators) we've recorded a CD titled HUMANOID that features the Prosthetic Head doing the vocals – reciting poetry and making singing sounds. There has also been a 'Fractal Remix' of HUMANOID. Incorporating the research of Cameron Jones, from Swinburne University in Melbourne, fractal shapes are thermally printed on the data surface of the CD. Every CD behaves differently depending on the specific printed shape, its colour and location on the CD surface as well as the user's own hardware and software system. The duration of each track will vary depending on the remix process.

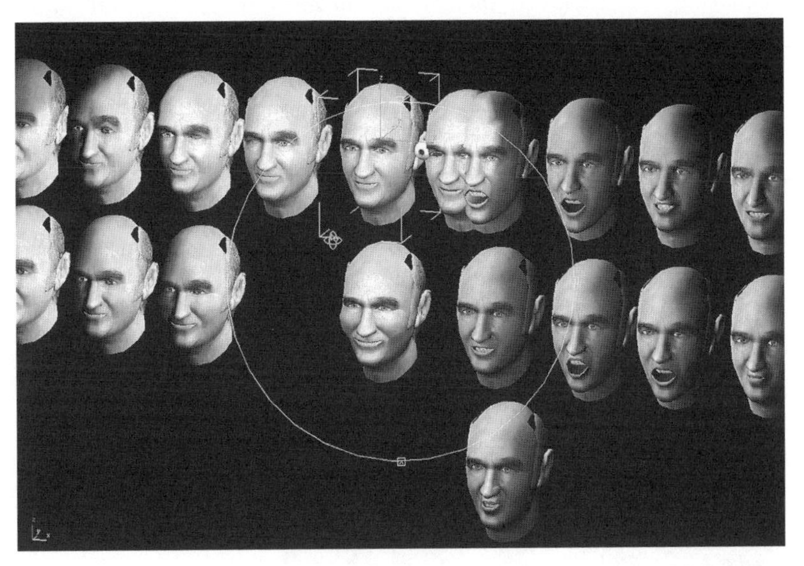

Figure 5.3 CLONED HEADS, San Francisco 2002. Image: Barrett Fox. © STELARC

Since the beginning of the Thinking Head project, a number of improvements have been made to the Prosthetic Head by Martin Leurssen and Trent Lewis at Flinders University, Adelaide. Mistakes in its AIML programming have been cleaned up. It is now possible to script particular facial expressions and emotions with a particular verbal response. The duration of pauses within a response can now be specified, allowing more expressive and poetic sounding responses. Additional plug-ins have been added to extend its repertoire of responses. The Head can now do head tracking thus being able to lock into the user's gaze, both with its eye movements and head turning. This creates a more seductive relationship between the Head and the user. A new modular software architecture is now being developed and implemented to facilitate the plugging in of new modules of capabilities. Speech recognition becomes a necessary interface to pursue. Perhaps ASR will be more successful in combination with lip-reading software. The agent needs to be more responsive to the user's appearance, attitude and emotional state. A more empathic and appropriately emotional agent will relate better with the user.

In listening to Catherine Pelachaud's plenary address, titled 'Virtual Humanoid as an Expressive Human-machine Interface', at HCSNet 08 at the University of New South Wales, it struck me that to amplify communication and enhance the expression of emotion and empathy with the user, it would be advantageous to use the virtual camera. We have a virtual human in a virtual environment. Why not use a more cinematic approach to how the

Figure 5.4 SECOND LIFE AVATAR, Melbourne 2008. Image: Daniel Mounsey. © STELARC

user experiences the virtual human? A virtual camera engine could drive how the virtual human would appear to the user. In other words the virtual camera views would be mapped to the virtual human's intent and affect. The virtual camera's switching and mixing of edits would be choreographed by the virtual human's behavior. The virtual human is seen as it wants itself to be seen. Or how it wants its subtle behavioral cues to be noticed or amplified. This would be especially useful in interacting with a full bodied agent on screens smaller than human-size when a glance or a nod of the head or a finger gesture might be missed.

WE NOW LIVE IN STRANGE TIMES. OF HAVING TO FUNCTION WITH MIXED AND AUGMENTED REALITIES AND WITH MULTIPLE EMBODIMENTS. HUMAN FLESH AND FLUIDS CIRCULATE BETWEEN BODIES. BLOOD IS EXTRACTED AND INSERTED INTO OTHER BODIES. ORGANS ARE DISPLACED, LIMBS ARE DETACHED, FROZEN SPERM IS THAWED AND FERTILIZES FEMALE EGGS IN VITRO THAT ARE RE-IMPLANTED IN A THIRD BODY TO ENABLE BIRTH. DEAD BODIES NEED NOT DECOMPOSE (PLASTINATED) WHILST SIMULTANEOUSLY COMATOSE BODIES NEED NOT DIE (SUSTAINED ON LIFE SUPPORT SYSTEMS, STILL BEATING AND BREATHING) AND CRYOGENIC BODIES AWAIT REANIMATION. CHIMERAS ARE CREATED. ALTERNATE ANATOMICAL ARCHITECTURES ARE ENGINEERED. THE DEAD, THE NOT YET DEAD, THE YET TO BE BORN AND THE PARTIALLY LIVING ALL NOW COEXIST IN SHARED TECHNOLOGICAL AND ELECTRONIC SPACES. TO BETTER PERFORM IN THE TECHNOLOGICAL TERRAIN. BIOLOGICAL BODIES ARE PASSIFIED, PROMPTED AND PROPELLED BY MACHINES TO BETTER PERFORM WITH PRECISION AND SPEED. TO BETTER PEFORM, BIOLOGICAL BODIES NEED TO FUNCTION WITH MIXED AND AUGMENTED REALITIES TO INTERFACE BOTH PHYSICAL AND VIRTUAL ENVIRONMENTS IN LOCAL AND REMOTE SPACES. THE BODY INCREASINGLY HAS TO EFFECTIVELY INTERFACE WITH AVATARS AND INTELLIGENT AGENTS. IN FACT ONE CYBORG CONSTRUCT IS THAT THE REALM OF THE POST-HUMAN WILL NOT RESIDE IN THE REALM OF BODIES AND MACHINES BUT RATHER IN THE REALM OF INTELLIGENT AND VIRAL ENTITIES, MANIFESTED AS IMAGES AND SUSTAINED IN ELECTRONIC MEDIA AND THE INTERNET. BODIES AND MACHINES ARE PONDEROUS AND PERFORM IN GRAVITY WITH MASS AND FRICTION. IMAGES ARE EPHEMERAL AND PERFORM AT THE SPEED OF LIGHT. AVATARS HAVE NO ORGANS ...

Acknowledgements

The Prosthetic Head was originally completed by:

Karen Marcelo	Project coordination, system configuration, alicebot customization.
Sam Trychin	Customization of 3D animation and text to speech software.
Barrett Fox	3D modelling and animation.
John Waters	System configuration and technical advice
Dr Richard Wallace	Creator of alicebot and AIML. Alicebot advisor. Alicebot is a natural language artificial intelligence chat robot.

The THINKING HEAD PROJECT involves:

From Talking Heads to Thinking Head Heads: A Research Platform for Human Communication Science is a 5-year Thinking Systems Initiative project jointly funded by the Australian Research Council, and the National Health and Medical Research Council (ARC/NH&MRC). The project is led by Professor Denis Burnham at MARCS Auditory Laboratories at the University of Western Sydney (UWS); involves over 20 researchers from computer science, engineering, language technology, cognitive science and performance art at the University of Western Sydney, RMIT University, Macquarie University, Flinders University, University of Canberra, Carnegie Mellon University, the Technical University of Denmark, and Berlin University of Technology; draws upon the resources and methodological approaches of researchers in the Australian Research Council Network in Human Communication Science (HCSNet); and is spearheaded in the performance domain by Senior Research Fellow and UWS Artist-in-Residence, Stelarc. The dual aims of the project are to build a next-generation talking head via integration of contributions from computing, human-head interaction, evaluation, and performance teams, and to establish a sustainable research platform within which a myriad of research questions can be addressed.

References

Nietzsche, Friedrich. (1989). *On The Genealogy of Morals; Ecce Homo*. Trans. Walter Kaufmann. New York: Vintage Books.

Wittgenstein, Ludwig. (1997). 'Excerpt from *The Blue and Brown Books*', in *Readings in the Philosophy of Language*. Ed. Peter Ludlow. Michigan: MIT Press.

6

Performative (Dis)closures – Sensual Readings and Writings of the Positive Body

Paul Woodward

A tale of two conferences

October 2004. I am at the University of East Anglia, delivering a paper at the *In Yer Face Theatre Conference* with my colleague, Dr Josephine Machon. Our paper is entitled: *Tracing the (Syn)aesthetic in Sarah Kane's 4:48 Psychosis*. I begin with the words:

> *Due to a series of personal incidents in my private life which led me to look deeply at issues of mortality and depression and the possibility of living in the darker places of human experience, I felt uniquely drawn to Sarah Kane's last text ... it was as if I had travelled through fire and now felt strong enough to tackle such dark and beautiful subject material.*

My heart beating. Its obvious what I am talking about isn't it? The room seems sticky, tense, unsettled. I proceed to talk alongside Dr Machon of a production I directed of Kane's final text. A production filled with images of the 'underneath'. A set of soil from which grows a white carcass of a bed, like a flower in winter. Performers buried under the soil, for much of the performance, only to burst through to sprout arms and write on others bodies. So much body text, words that can be written yet not spoken. So many pauses. So many silences. A production that wants to say something and yet passionately does not say it. This is not just non-disclosure. This is dis(closure). A resolute shutting down of signification coupled with a desire to communicate yet thwarted, muffled, silenced loudly.

October 2005. One year later. I am at the University of East Anglia again, this time for the Jean Genet Symposium. I am halfway through delivering a paper entitled: *Tracing the Genetian Inheritance in Contemporary Body Art.* It is just me this time and I am feeling it. I am arguing that Genet's representation of the abject body articulates a stunningly original technique of transgression that contemporary artists, such as Ron Athey and Franko B., have drawn upon and developed further in their respective practice. In the last instance I talk explicitly how I experience a form of cleansing by proxy when confronted with the spectacle of Franko B.'s blood draining away from him on a vast white canvas in his 2001 piece *Oh Lover Boy*. Sweat, large amounts of it, trickling down the underside of my shirt as I deliver my paper. This pausing sweating disorientated academic. Here it comes. It is as if the words are on autocue coming towards me and now I am reading the words of my script and I am pausing and palpitating at the lectern. Dizzy, somewhat outside of myself. The words exit as if not only from my mouth and fill the lecture auditorium:

> *My outcast, abject body felt cleansed. The trauma of my own HIV positive diagnosis — the fear of decay, the pain of illness — suddenly drained away. I had moved from morbidity to affirmation. My doctors can employ the medical discourses to help me rationalize my body's battle with the HIV virus, yet it was my encounter with the live ritual of Franko B.'s bleeding body, which triggered a non-rational and transformative response. This could only occur within the transactions created by body art, primarily, because it is able, through the liveness of the body, to challenge discursive meaning.*

My first public disclosure moment. All wrapped up in reference and context and neat academic speak. But there it was all the same. It is out now. I cannot take it back. And I am left wondering what happened in that moment. Something happened, something changed, something shifted. An event of words and body, as much through my body as it was about my body. And I want to find the words to help describe it. The room is not sticky and tense, it seems somehow relieved. This is not disclosure. This is (dis)closure. I am in control I am inviting signification. I am inviting framing.

Two differing contexts, two differing disclosure events, each attempting to articulate the same thing. An experience of the body somehow forged into the word. Sometimes said directly, sometimes indirectly. But each time an invitation is offered to read the speaking body both literally and metaphorically as a positive body. And in each instance something has changed. Through the failure or success to articulate the words 'I am HIV positive' something has changed in the world. This chapter attempts to look at what might happen in this disclosure moment; when the invisible world of the positive body breaks forth into the realm of the visible.

Ethics, responsibility and wounded storytelling

In his groundbreaking and influential text *The Wounded Storyteller: Body, Illness & Ethics,* Arthur W. Frank suggests that *everyone* will experience an affliction of the body at some point in their lives, and as such will occupy the positioning of a 'wounded citizen' (see Frank, 1995). As such an entity, one may well aid the process of healing through the ritual of storytelling. The crucial act, according to Frank is to 'find one's voice' which the experience of being ill and its ensuing treatment has conspired to take away. This concept of 'the voice' in his writing embraces both the somatic and the esoteric following his assertion that the wounded citizen has been hurt in body, voice and spirit. The 'voice', whilst being a physical component of the body also 'speaks the mind and expresses the spirit'. The 'mystery of illness', in Frank's account, is best articulated through the body as a medium of expression as both source and object. In the 'silences between words', Frank opines, 'the tissues speak' (ibid.: xii)

Furthering his argument for an embodied knowledge, Frank suggests that in 'wounded storytelling' there is an inevitable intersection with the ethical. In allowing the' tissues' (ibid.) to speak of a highly personal subjective experience of the body, the storyteller not only seeks to reclaim the experience of suffering (through the alchemy of testimony), but simultaneously engages in a moral action:

> Kierkegaard wrote of the ethical person as editor of his life: to tell one's life is to assume responsibility for that life. This responsibility expands. In stories, the teller not only recovers her voice; she becomes a witness to the conditions that robs others of their voices. When any person recovers his voice, many people begin to speak through that story.
>
> (Frank, 1995: xiii)

If such stories involve an act of disclosure of HIV status, the notion of 'responsibility' evoked here by Frank takes on an even greater significance. The most prominent public health recommendation for people with HIV is to disclose or tell others about their diagnosis, especially their sexual partners (Burris, 2001; Rothenberg and Paskey, 1995). This urge to stimulate an individuals 'civic duty' as regards their openness about their diagnosis has been a major component of the public health response to the HIV epidemic. Yet this is only one aspect many HIV positive people experience as part of their 'duty' as 'wounded citizens'. To inform others of their health status who might be susceptible to transmission may form the first, and most immediate line of ethical action but this is often accompanied by an equally strong sense of purpose to make sure others become knowledgeable about the disease. In such scenarios there is a direct motion to recognize the critical importance of respecting those who are living positive lives. Frank similarly writes of the wounded storyteller as s/he who also embraces the dual role of the wounded healer – a pathway which he articulates as his most significant journey through his own illness narrative.

In both the disclosure examples I gave previously, I am reminded of the palpable sense of 'danger' and 'risk' that these events evoke. There is a tension and subtle interplay here between private and public information. When people consider disclosure of HIV status, they are often dealing with a decision to allow access to the information to some(one) and at the same time to deny access to others. The idea of 'private information' relates to information based on fact and feelings to which others would not normally have access and so exists in a constructed domain separated from the 'public'. People do not indiscriminately reveal public information, however, because doing so would make them feel too vulnerable (see Gilbert, 1976; Petronio, 2002). People more than likely calculate how much they want to tell, when they want to tell, and who they want to tell for the very reason that the information is risky (Petronio, 1991, 2002).

> People are social beings with needs to connect as well as needs to separate from others. In many ways, this is the paradox of the HIV dilemma. There are conditions that justify withholding information about a person's HIV status from others .... Yet, to obtain the much needed social support or because others may be affected, disclosure is necessary. The key to navigating the markers between private lives and shared ones is people's decisions to open up completely, partially, or keep their privacy boundaries closed.
>
> (Greene, Derlega, Yep and Petrinio, 2003: 17)

I was acutely aware at the time of my disclosure at the Genet Symposium that I was at that time not fully disclosed to my own parents. But once the 'truth is out there', there is very little control over that information. It seems that many people are fully disclosed in private but not so in public, and so live in a strange world of limbo in which aspects of dis(closure) and (dis)closure exist simultaneously.

What happens at the moment of disclosure?

What one is witness to in the disclosure event, whether full, partial or restricted, is primarily a speech utterance. One that is distinct in that it is at once both an act of speech and an act in itself. The sequence of words 'I am HIV positive' can be seen to describe both a medical condition and at the same time serves to signify the narrative event of diagnosis through to some level of acceptance. Concurrent with this revelation is a transgression. What is transgressed is the personal boundaries a person might erect for protection's sake. The intertwined privacy and disclosure process usually consists of at least two individuals engaged in a social interaction, each with his or her own feelings, beliefs, attitudes, values and expectations, and the behaviour of both persons is affected by the social, psychological, relational and physical context. Such a disclosure of personal or private information is shared by one person with another specific person or persons and in this sense self-disclosure distinguishes itself from public disclosure. And yet in my experience of disclosing my status, despite these differences, there is at the same time a sense of sameness – each time the words are spoken something changes, more than this the experience of telling remains the same regardless of context or indeed, audience. The very utterance of the words 'I am HIV positive', then, move beyond mere statement or description into a whole other linguistic realm in which these words are in themselves actions which have the potential to 'do' something in the world.

Performativity and disclosure

The English academic J. L. Austin coined the word 'performative' in the first instance which points up the way in which our utterances can be *performative*: words doing something in the world. To *say* something is to *do* something. In uttering certain sentences people perform acts. Austin suggests that everyday people in everyday situations and places use a variety of such utterances. Promises, assertions, bets, threats, curses, judgements and thanks are, according to Austin, actions of a distinctively linguistic kind in the first place, but also they are actions in themselves. Austin 'argued that words are not purely reflective, that linguistic acts don't simply reflect a world but that speech actually has the power to *make* a world' (Jackson, 2004: 2). In other words they do not describe actions: they are actions. They are 'performed', like other worldly events and thus make a difference in the world:

> The term 'performative' is derived, of course from 'perform': it indicates that the issuing of the utterance is the performing of an action. The uttering of the words is, indeed, usually a, or even the, leading incident in the performance of the act.
>
> (Austin, 1975: 8)

Austin broadened his basic outline of the performative into a theory of 'speech acts', in which the performativity of requests, orders, and declarations etc, were seen to characterize all the utterances we issue as speakers and as such are a part of everyday life (Schechner, 2002: 122). The words themselves, however, need to be corroborated by an apposite action – the speech act becomes both precursor and inciter of a physical act. Austin himself acknowledged the complexities of codependency that performatives exhibited with incited and apposite actions and saw that this in itself could be seen as pointing to a weakness in the power or 'force' of performative utterances themselves. In order to address this he made a series of distinctions with which he could classify the effectiveness of utterance and physical action. He reasoned that performatives uttered under false circumstances, for instance, were 'unhappy' or 'infelicitous' (Austin, 1975: 14).

The disclosure performative

Such an event is, in itself, of profound import when considering the nature of a health disclosure event such as stating that one is HIV positive. Released from the simple declarative constraints of the realm of the constative and in obeying the law of the descriptive fallacy, such an utterance could be seen to both making and remaking the world at the moment of its production. In this way the utterance 'I am HIV positive' belongs to the world of such explicit performative utterances favoured by Austin himself, such as 'I do' at a wedding service, or 'I love you' in a romantic engagement or even ' I name this ship' – all essentially three word linguistic events which straddle both the constative and the performative simultaneously. How a HIV disclosure becomes a performative event, however, means that a whole system of linguistic, social, political and performance related issues come to the fore.

What is interesting here is that a disclosure rarely takes the form of 'I *have* HIV', but instead is most often to be articulated by the utterance 'I *am* HIV positive'. This begs further analysis as to why it is the case as we tend not to say – in the case of cancer, for example – 'I *am* cancer'. It would seem, therefore, that there is something rather complex occurring in this utterance which may suggest a range of interpellative processes. Again, the act of speaking enacts a series of doing, in which a range of discourses begin to invade the speaking subject.

To return to Austin's prime example, the 'I do' in a marriage is of course in reference to 'taking the hand of x to be my lawfully married wife/husband' whereas the 'I am' in a disclosure scenario is ambiguous in terms of the relational qualities of, on the one hand, the speaker and, on the other, a virus named HIV. In Austin's prime example the act of saying is the 'doing' or 'legal binding' between two consenting people, witnessed in the 'calm interpellation' of the congregation/state – a union classically sealed with a kiss (or 'x') at its end. In my prime example of a disclosure scenario, the act of claiming/naming is creating a binding of sorts between self and virus till they become indivisible – the witness to this speech act, then, in some ways becomes complicit in the union between 'self' and 'virus' and sealed with a '+'. What is being performed here is a representation of a union at the level of body created through an act of speech which makes it reality. A wedding of sorts – maybe what should be said as part of a public disclosure scenario is, 'I do take this virus to be part of my body, for better or for worse, for richer or for poorer, till death do we part'. So if the speech act is creating in the world a union in this manner, what is being displaced here?

Dis(closure) – theatres of concealment

It is impossible to ignore the factor of stigma as risk criteria which influence disclosure decisions. The sociologist Erving Goffman identifies stigma as that which 'refers to an attribute that is deeply discrediting' (Goffman, 1990b: 3). It is the mechanism by which someone is seen as being 'spoilt' or 'tainted' , or making someone seem inferior in the eyes of others partly because s/he may fail to live up to others expectations. A newly diagnosed HIV positive person can feel the force of this stigma in two ways, firstly in an internalized manner and, secondly, as an externalized 'tainting' as they experience the reactions of others post-disclosure.

According to Goffman, a distinction can be made between two types of stigmatizable persons. A 'discredited' person assumes that a supposed undesirable characteristic is visible and known to others. The person's task is to somehow manage the tension associated with having this information known. A 'discreditable' person assumes that the undesirable characteristic is not visible or known by others. For this person, s/he may attempt to conceal the information from others or to 'pass'. For the undisclosed HIV positive person, Goffman's 'type' of a 'discreditable' could be seen to apply. The secrecy involved and the pressures of 'keeping it all in' casts the subject in a particular frame of internal/invisible. The 'discreditable person' who wants close relationships may decide to disclose the information to some people but not to others, or the discreditable person may decide to tailor how much to tell about themself to reduce the possibility of being rejected or hurting others (Goffman, 1990b: 38):

> By hint or sign, many sought to convey the possibility of infection in a way that avoided embarrassment and limited the humiliation and rejection that could result from more explicit declarations. A partner's response might then lead to further discussion or elucidation. But as individuals sought to 'speak' indirectly, or without words, unclear signs had to be deciphered. Thus, in the grey zone between deception – with all its moral ambiguity – and truth telling, codes flourished.
>
> (Klitzman and Bayer, 2003: 46–7)

What Klitzman and Bayer seem to be highlighting here is that in concealment scenarios there is a distinction made between 'lying' and 'deceiving', and that each are played out against each other as these codified systems are played out in both public and private. I am interested in how such systems are 'performed', both textually and subtextually, and to what extent these 'theatres of concealment' can be identified in everyday life and in contemporary performance strategies.

In thinking about how one might develop a 'poetics of dis(closure)', I am compelled to reflect on my own body of directed/devised work created with undergraduate students after my own HIV+ diagnosis in 2000. I am struck by a body of repeated core themes and images which seemed to exist under the apparent surface of this collaborative work. In many ways the very titles of these projects themselves seem to be articulating some thing of the 'discredited' identity – *4:48 Psychosis* (2002); *Unspeakable* (2003); *The Icarus Factor* (2004); *A Wounded Place* (2005). Starting with the emotional extremities of Sarah Kane's text of *4:48 Psychosis*, I now identify my own sense of shock and numbness as being present there. In retrospect they all seem to be saying something in code. I was trying to say something publicly that I could not, or simply was not ready to, say out loud, in its purest terms. I began to wonder to what extent these devised productions contained codified expressions of my diagnosis and journey towards acceptance. And if they did, then could they be seen to display ritualistic tendencies of 'identity transformation' or 'augmentation' (Turner, 1982). I seemed to be learning a language of disclosure that only really found its true exposition in last year's performance project *The Healing Room* (2007).

It was this possibility that prompted me to investigate where I may have been performing acts of self-disclosure both on and off stage. Victor Turner (1982), suggests that human beings in a state of flux, or 'becoming', are often found to exhibit tendencies to appropriate artistic material for their own transformative ends. In some sense I now see that, working within the devising paradigm, I was opportunistically reaching out to the medium of theatre for a form and content – for a form of disclosure of self to self. I have come to believe that I was seeking the ritual effects of these codified disclosures as performance in an effort to strip away my sense of internalized stigma surrounding my positive diagnosis. Or, as Erving Goffman suggests, I was attempting to enact a ritualized 'stripping away' of an 'accumulated, acculturated sense of abnormality' (Goffman, 1990b).

I was very much inspired by the ex-catholic priest David Garrick who asserts:

> Coming out [is] a paradigm for all performances of self: performances in which one seeks – for whatever reason – to unpack, to put into words or symbolic gestures, to reveal unmistakably, to disclose, the truth about oneself to witnesses.
>
> (2001: 11)

Following Turner, Garrick suggests that an orchestrated coming out has the processional structure of the 'rite of passage' that creates new identities. Though its boundaries are often blurred in practice, the rite has three discernable phases: (1) a separation phase of preparation and setting things in motion; (2) a transition phase of the self-disclosure itself along with the witnesses response; (3) a reincorporation phase, when the performer returns to everyday life with the new identity in place. Placing this within the narrative structures of a dis(closure) scenario brings me one step closer to defining what might define such an event.

Towards a definition of dis(closure)

- The processes of concealment by which a positive subject might inadvertently perform highly complex and ritualized acts.
- The subject, in attempting to close down interpretation of sero-status, inadvertently sets up a whole range of codified behaviours.
- This could result in identifiable tendencies in the process, presentation and reception of performance work.

(Dis)closure – theatres of revelation

A fully disclosed HIV positive person, Goffman's 'type' of a 'discredited', could be a state in which the forms of (dis)closure could be said to apply – the 'undesirable characteristic' being out in the open, it can now be seen to operate in a particular frame of the external/visible. Of course the speaking subject here, through the vehicle of a speech utterance, has made themselves extraordinarily vulnerable as they are seen to 'inhabit' this 'undesirable characteristic' and possibly vice versa.

In such an instance we can see how Frank's invocation of Kierkegaard's will to assume responsibility 'expands' as the speaking subject, prone to the power of stigmatization, becomes as if an editor of his/her own life. If the primary and secondary features of (dis)closure are not to be seen to 'rob both self and others of their voices', then the recovery of that voice casts the speaking subject as both witness and testament to a 'discredited' life. If, as Frank's argues, to speak of 'one's life is to assume responsibility for that life', then it is at the site of this first utterance in which the speaking subject claims a positioning from which the possibility arises of 'many people begin*(ing)* to speak through that story' (Frank, 1995: xiii).

So what might this (dis)closed model look like in terms of performance practice? I am very much moved by the solo work of the celebrated Wooster Group core member Ron Vawter in this respect. In 1995 he performed his show *Roy Cohn/Jack Smith* in various venues across America. As a two-part show, Vawter portrayed two historical figures who have nothing in common except their homosexuality and their death from AIDS in the late 1980s. Within the piece the politician Ron Cohen and the performer Jack Smith are positioned as polar opposites of gay lifestyle and sensibility. Both figures, however, suffered from a life in the closet as a result of their formative years in the repressive and punishing environment of 1950s America.

What I find most interesting about Vawter's performance is the way in which, despite drawing his characters in great detail, he also gave access to his own personality to be present on stage simultaneously, through which a critical distance was observed. Crucial to this process was his prologue to the show in which he discloses his own health status, announcing that he is a person living with AIDS, which acted as a framing device for the whole show. In his book *Acts of Intervention: Performance, Gay Culture, and AIDS*, David Roman observes how Vawter's disclosure is key to the shows performance apparatus:

> Vawter challenges the interpretive logic of his performance, which links repression with AIDS, within a causal relation by the use of his own body and experience with AIDS. He manipulates interpretation through this experiential performative.
>
> (1998: 138)

Through his primary disclosure Vawter invited his audiences to observe his body as 'discredited', yet in doing so is seen to be in control of its signifying processes. Whilst he himself exists corporeally in the realm of the (dis)closed, his characters parading upon the stage are seen to perform in the realm of the dis(closed). What the audience were present to was a ritualized act of resistance, witnessing an interplay between these two states of being.

The English playwright David Carter's 1995 piece *Viral Sutra* offers a similar example of the potential for the (dis)closed mode of making/reading performance. In the text Carter writes about not just any virus, but *his* virus and his bodies relationship to it. He takes us inside his host body to stage the moment of sero-conversion itself. Whilst the body of Carter is offered as a living organism, it is his submicroscopic entities CAP, GAG and POL who work through what it means to be caught up in an endless process of viral replication. In staging his own body's story of sero-conversion we see the act of imagination and dramatic explication become an act of resistance in itself, a reclaiming of the body just at the moment when the body loses some of its autonomy to the invading 'other' of a virus that refuses to be called 'it'. Released from the cat and mouse game of 'is he isn't he' in terms of sero-status, Carter's body becomes text itself in the (dis)closed mode and recognizes its resistance to closure of serostatus reading so that we read beyond such frames and become conscious of emer-ging discourses affecting the self at the moment of revelation – just as Carter's imaginary submicroscopic entities attempt to redefine their new identities in a state of viral flux and biological transformation.

Taking inspiration from such works as Vawter and Carter's and on reflection of my own dis(closed) tendencies displayed in my own production work post-diagnosis, I now look upon such work differently. I seemed to be learning a language of (dis)closure that only really found its true exposition in last years performance project *The Healing Room* (2007). Again, it is all in the title. For this production I resolved to be fully disclosed from start to finish, and in being so observed how this affected both the process and product. This production I see now almost as an antithesis of my earlier work on Kane's text *4:48 Psychosis* – if that production was all about the 'underneath', this production was about the 'beside' or even the 'beyond' as I sought to find an aesthetic for (dis)closure practice. This series of 'rooms' held under a proscenium arch offered a healing space which dealt with positive scenarios both explicitly and implicitly throughout using a mix of (syn)aesthetic text (see Machon, 2009), ritual and movement. Interestingly, although dealing with explicit experiences of the body positive, both performers and audiences created their own pathways through the body of work in terms of offering their own body stories as an integral point of integration in both process and reception of product. I conclude this chapter, then, with a working definition of my proposed performance model.

Towards a definition of (dis)closure

- Recognizes its resistance to closure of sero-status reading.
- The subject is seen to be in control of both bodily and speech signifying processes.
- The (dis)closed subject acknowledges the efficacy of performance.
- Is conscious of emerging discourses affecting of the self at the moment of revelation.
- Challenges the stigmatized characteristics of being an openly positive person in society.
- In doing so offers a window of opportunity to engage in ritualized acts of resistance.

References

Austin, J. L. (1975 [1962]). *How to do Things with Words*. Oxford and New York: Oxford University Press.

Burris, S. (2001). 'Clinical decision making in the shadow of law', in *Ethics in HIV-Related Psychotherapy: Clinical Decision Making in Complex Cases*. Ed. J. R. Anderson and B. Barret. Washington, DC: American Psychological Association Press.

Carter, D. (1995). *Viral Sutra*. Finborough Theatre. London

Finburgh, C., C. Lavery and M. Shevtsova (eds). (2006). *Jean Genet: Performance and Politics*. Basingstoke and New York: Palgrave Macmillan.

Frank, A. (1995). *The Wounded Storyteller: Body, Illness, and Ethics*. Chicago and London: University of Chicago Press.

Garrick, D. (2001). 'Performances of Self Disclosure: A Personal History', *The Drama Review*, 45.4 (T172), Winter (New York Uni & Massachusetts Institute of Technology).

Gilbert, S. J. (1976). 'Empirical and theoretical extensions of self-disclosure', in *Explorations in Interpersonal Communication*. Ed. G. R. Miller. Beverly Hills, CA: Sage.
Goffman, E. (1990a [1959]). *The Presentation of Self in Everyday Life*. London and New York: Penguin Books.
——. (1990b [1963]). *Stigma: Notes on the Management of Spoiled Identity*. London and New York: Penguin Books.
Greene, K., V. Derlega, G. Yep and S. Petronio. (2003). *Privacy and Disclosure of HIV in Interpersonal Relationships: A Sourcebook for Researchers and Practitioners*. New Jersey and London: Lawrence Erlbaum.
Jackson, S. (2004). *Professing Performance: Theatre in the Academy from Philology to Performativity*. Cambridge: Cambridge University Press.
Petronio, S. (1991). 'Communication boundary management: A theoretical model of managing disclosure of private information between marital couples', *Communication Theory, 1*. Albany: State University of New York Press.
——. (2002). *Boundaries of Privacy: Dialectics of Disclosure*. Albany: State University of New York Press.
Klitzman, R. and R. Bayer. (2003). *Mortal Secrets: Truth and Lies in the Age of AIDS*. Baltimore, MD: The Johns Hopkins University Press.
Kosofsky Sedgwick, E. (2003). *Touching Feeling: Affect, Pedagogy, Performativity*. Durham, NC, and London: Duke University Press.
Loxley, J. (2007). *Performativity*. London and New York: Routledge.
Machon, Josephine. (2009). *(Syn)aesthetics – Redefining Visceral Performance*. Basingstoke and New York: Palgrave Macmillan.
Parker, A. and E. Kosofsky-Sedgwick. (1995). *Performativity & Performance*. London and New York: Routledge.
Roman, R. (1998). *Acts of Intervention: Performance, Gay Culture, and AIDS*. Bloominton and Indianapolis: Indiana University Press.
Rothenberg and Paskey, (1995).
Schechner, R. (2003 [1988]). *Performance Theory*. London and New York: Routledge.
——. (2006 [2002]). *Performance Studies : An Introduction*. London and New York: Routledge.
Turner, V. (1969). *The Ritual Process: Structure and Anti-Structure*. Chicago: Aldine de Gruyter.
——. (1982). *From Ritual to Theatre: The Human Seriousness of Play*. London and New York: PAJ Publications.
Vawter, R. (1995) *Roy Cohn/Jack Smith*. Performed in various venues across America.
Woodward, P. (directed and devised works) (2000). *4:48 Psychosis* by Sarah Kane. St.Mary's University College, Strawberry Hill, London.
——. (2003). *The Icarus Factor*. St. Mary's University College. Strawberry Hill, London.
——. (2004). *A Wounded Place*. St. Mary's University College. Strawberry Hill, London.
——. (2006). *The Healing Room*. St. Mary's University College. Strawberry Hill, London.
Woodward, Paul and Josephine Machon. (2002). 'Kane & (Syn)aesthetics – Tracing the Experiential in the Work of Sarah Kane', *In Yer-Face? British Drama in the 1990s*. Conference, University of the West of England, Bristol, 7 September.

7

The Supernatural Embodied Text: Creating *Moj of the Antarctic* with the Living and the Dead

Mojisola Adebayo

Moj of the Antarctic: An African Odyssey is a physical storytelling theatre piece for one performer with poetry, music, dance, audience interaction and visuals.[1] The play tells the odyssey of Moj, a woman who escapes slavery in the deep south of America by cross-dressing as a white man, travels to England, becomes a sailor on board a whaling ship bound for the southern ocean, and becomes the first African woman to step foot on Antarctica. The play is inspired by Ellen Craft, a mid-nineteenth-century African-American woman who in 1848 actually escaped slavery by cross-dressing as a white man.[2] *Moj of the Antarctic* is an intertextual fusion of Ellen's real life boundary-breaking trans-gender, trans-racial, trans-geographical performance with the voices of almost 20 dead authors, as well as digital images and film of myself performing as 'Moj', shot on location on Antarctica by legendary Queer photographer and film-maker, Del LaGrace Volcano.

In this chapter I will describe some of my creative process in making this performance text and will discuss how the experience of creating, writing and performing, at times, feels supernatural; as if the dead are present in the living work. I will demonstrate how Ellen Craft's biography merges with my imagination to become a new fiction for the stage and in turn shapes my own story; my 'auto-biography'. History becomes story and once again history. In outlining this Marxian cycle of thesis – antithesis – synthesis, that is, biography – fiction – autobiography, I will begin at the end, with a little of my autobiography.

I grew up as a Black girl of mixed Danish and Nigerian (Yoruba) cultural heritage in London in the 1970s. The only time I saw other Black people represented in performance was on popular television situation comedy shows such as *Till Death Us do Part, Mixed Blessings* and *Mind Your Language*. Although these programmes affirmed our very existence within British society and went some way to satisfying our need to be recognized in a white environment, which was, quite literally, ignorant of us, my feelings towards these programmes was, as Homi Bhabha has discussed, ambivalent. The scripts appeared to ridicule us as well as represent us, to mimic us and to mock us.[3] These were sit-coms where the joke was on us. If, as Stuart Hall has discussed, representation produces meaning, for me these 'sit-coms' were meant at worst to reinforce pervasive racist ideologies and at best to convey the idea that we were foolish foreigners who were out of place in Britain.[4]

However, one extraordinary television serial screened in 1977 was responsible for breaking this monotony, for representing Black history and Black people from a Black perspective, thereby bringing new meaning to a generation of Britons of all races. This was the African-American eight-part television serial drama *Roots*, based on the Pulitzer prize-winning novel by Alex Haley.[5]

In researching and writing *Roots*, Alex Haley traced his family history from the United States of America back to what is now The Gambia, in West Africa. Through history narrated by a Griot (a West African oral historian and bard), Haley traced his roots in Africa back to a young man who was captured, enslaved, brought to America and forced to take the name 'Toby'. His name was Kunta Kinte. The televised version was the first time I had seen so many Black people represented through performance. It affected not just me, but influenced a whole generation. One successful Black British playwright recounts watching *Roots* as a child:

> It was a moment that changed my life. By the end of the series I had told my mother that I would one day trace my heritage back to Africa and reclaim an ancestral name. Before I watched the programme I was called Ian Roberts but now my name is Kwame Kwei-Armah.[6]

Roots instilled in many of us an enormous sense of Black pride. The *Roots* experience was the first time I perceived that performance had the power to increase understanding, change perspective and empower a whole people. Through merging lost history with his imagination to create a new work of fiction, Haley discovered who he was in the world, *Roots* therefore is an essential part of his own autobiography.

Roots was not only a great source of inspiration and information to me in its description of the history and legacies of the trans-Atlantic slave trade, it also provided me with a model in writing *Moj of the*

Antarctic. Following Haley's trajectory, I work from biography to fiction and in turn the process of making this work has become part of my own autobiography, the story of my life.

The trans-Atlantic slave trade may be considered an act of genocide. The term genocide literally means race (genus) killing (cide).[7] We often think of genocide in relation to the attempt and actual murder of an entire ethnic population such as the Jewish people by the German Nazis or the Armenian people by the Ottoman Empire. However, to kill a race is not only to kill it physically, but to kill it mentally. To wipe out its people's names, culture, languages, traditions, religions, identities, histories and stories. In most acts of genocide the perpetrators of the crime attempt to cover up the evidence, and erase histories. Therefore liberation struggles often attempt to rediscover and reclaim lost histories. This is also the experience of the African Diaspora.

One such lost history is the life of Ellen Craft. I have long been fascinated by the story of this amazing woman who had the courage and skill to perform the role of a white man in real life, in order to liberate herself (and her husband William Craft) from the crime against humanity that is slavery. I knew that I wanted one day to write a play inspired by Ellen Craft's experience, although I did not know where to start. I decided to take a trip to The Gambia for a holiday and to visit James Island, where Alex Haley's ancestor, Kunta Kinte, was once held captive.

James Island is an historic slave site where millions of enslaved Africans were held as cargo on their way to Europe and the Americas. To reach James Island we had to cross the river Gambia in a small boat. I have been to Belsen, one of many Nazi concentration camps where millions of Jewish people as well as disabled and deaf people, Gay people, trade unionists, Communists and Roma people were tortured and murdered. James Island is the only place I have ever been to which has a similar haunting atmosphere, you can almost smell death. We were told by the local guides that thousands of African people threw themselves into the crocodile infested river Gambia, and also into the Atlantic ocean, in order to escape slavery through the act of suicide. For them, liberation meant death. So many African souls sleep on the beds of the Gambia river and the Atlantic Ocean as a result of slavery, and sadly so many today are washed up on the shores of islands such as the Canaries, at the feet of European holiday-makers. These are 'illegal immigrants', people attempting to escape the poverty which is a direct legacy of the slave trade, perpetrated hundreds of years before. I am not a religious or superstitious person, but on that small boat in the river Gambia, I felt the presence of the spirits of ancestors lost. It is they whom I attempt, in part, to honour in this play.

On James Island we visited the cells in which Kunta Kinte, and many others, would have been held. I was told that many of the wooden carvings seen on the mainland replicate the depressed physical position that slaves sat in whilst waiting to be transported. The figures in the carvings sit on the ground, backs bent, with one hand on their head and the elbow resting on one knee, and the other leg stretched out long with a bowl at their feet. In *Moj of the Antarctic*, the first character I play is the Griot, much like the Griot who would have narrated Alex Haley's family history to him. She begins the performance by sprinkling a libation of water onto the ground and over the audience, in homage to the ancestors, and by speaking in 'tongues' – the mutterings and mumblings of her spirit. But even before I leave the dressing room in the role of the Griot, the first drop of water I pour before every performance is into the bowl of the African slave woman in the carving which I brought back from The Gambia. A reminder to remember, to never forget.

Upon sailing away from the island we were informed that in years to come James Island, a UNESCO World Heritage site, will disappear into the sea. This historic place will be lost due to the devastating effect of coastal erosion caused by rising sea levels, as a result of melting Polar ice caps, which is due to global warming. I had for some time been interested in the subject of the Polar regions and Antarctica in particular. I had wondered if I could make a piece of theatre that, imagistically, brought the continents of Africa and Antarctica together, in order to speak about climate change. In the little boat I remembered Ellen Craft. I decided then that I would make a play inspired by Ellen, about a woman called 'Moj' (my shortened name), who escapes slavery in America by cross-dressing as a white man, becomes a sailor on a whaling ship and journeys all the way to Antarctica. Through her the two continents, which were once joined as the super continent Gondwana, would be brought together again. Through her I could speak about Africa and Antarctica, black and white, female and male, the legacies of the slave trade and the legacies of the industrial revolution, rising seas and melting ice. The story would be introduced and concluded by a Griot, who would put the play into the context of climate change and its effect on Africa and Africans, before going on to guide us on Moj's odyssey.

When I returned from The Gambia I started making plans to travel to Antarctica to carry out research for the play and to shoot visuals for projection in the performance. With the synopsis in my mind, and support from Arts Council England, I took a research and development field trip to Antarctica in 2005 with legendary photographer and filmmaker of Queer sexualities and genders, Del LaGrace Volcano. Loaded down with mid-nineteenth-century male costumes, white face paint, black face paint and much-needed thermal underwear, I improvised as Moj, whilst Del shot the footage. I never imagined as a child that my own biography would read 'the first Black woman to perform on Antarctica', thanks to Ellen Craft.

Del and I had a very basic plan and knew that we would have to shoot quickly due to the cold and the sensitivities of this most delicate location. Although the landscape was awe-inspiring, we did not stop to reflect emotionally or spiritually on what we were doing, that is, until the day came when we planned that I would cover my face in black paint. This was to be a kind of inversion of Ellen's crossing as white, reclaiming the melanin. During my research I had seen photographs of the famous British Polar explorer's Captain Scott's men engaging in Black face minstrelsy for entertainment during expeditions to Antarctica a century ago.[8]

The discourse of Polar exploration is layered with the language of white supremacy. It is hardly surprising, as the era of conquest in the South Pole was a period synonymous with the colonization of Africa and Africans and the racist ideologies that justified the actions of European nations. I wanted to reference minstrelsy, a rarely mentioned feature of Polar history, I wanted to mimic their mimicry and mock their mockery. It was at the moment I sat on a snow covered black rock on Antarctica (Figure 7.1), looked into a mirror and smeared my face with black paint, that I started to weep. It felt that the tears were coming from a whole line of Africans going back through 400 years of humiliation and oppression. We were grieving. I decided to sit in the slave position of my Gambian wooden carving and allow black tears to flow on this snow-covered wilderness. I have never before felt the presence of the dead so close. But on that rock in Antarctica I felt that the millions who had perished in the sea and endured slavery were all around me. Del was utterly silent and continued to shoot the photographs that were to become an integral part of the live production. The conclusion of the play was 'written' in that live moment.

I returned to Europe full of ideas and Del's outstanding photography to inspire my writing. I knew that I would not be honouring the spirit of Ellen Craft if I did not now write the play, and the whole trip would be a waste of carbon and time. But I felt nervous and insecure. My

Figure 7.1 Mojisola Adebayo shot on location on Antarctica. Image © Del LaGrace Volcano, 2005

background was as a devisor in theatre, not a 'writer'. I had never before sat down at a computer to write a play without being in a rehearsal room with actors improvising the text. So I started to think of the writing task as a devising project. Only this time my co-devisors would be writers from the mid-nineteenth century. I read narratives written by African-American former slaves such as Harriet Jacobs, Harriet Wilson, Phillis Wheatley, Frances Ellen Harper, Frederick Douglass, as well as influential white writers from the same period, all of them deceased. I then sat at my computer and imagined they were all with me in a rehearsal room and I would weave their words, as well as my own, into the script. I made collages from sentences and played with the words just as I would with actors' improvisations, and even Karl Marx can be made to sound like a rapper:

> *Moj reads from* The Communist Manifesto:
> In my own century
> Nature faces subjection
> To man and machinery
> Chemistry and industry
> Railways, electricity
> Steam-navigation

Clearing whole continents
For cultivation
Conjured out of the ground
Are whole populations![9]

I had edited over 30 scripts from devised processes before. This time, for the first time, I was devising with the dead.

As well as mid-nineteenth-century writers I included fragments of William Shakespeare, taken from what might be considered the first English colonial play, *The Tempest*, fittingly a play which begins with the a man-made storm, anthropogenic climate change. At the point where Moj decides to escape slavery she begins to grab the air as if grabbing for some invisible freedom. Her grabbing hand becomes a fist. In her fist she sees the face of her grandmother and it is here she finds her inspiration. She says:

> My grandmother was an African. Born on a slave ship, in the Atlantic Ocean, birthed through the muted screams of a mother's lips bolted. Her father threw himself into the sea with chains about his feet, he would rather become a fossil than see his family treated like cattle.[10]

Then she begins to sing words from *The Tempest*, words which I slightly altered to fit my narrative:

Full fathom five
[my great-grandfather] lies
of his bones are coral made:
those are pearls that were his eyes,
Nothing of him that doth fade,
But doth suffer a sea-change,
Into something rich and strange ...[11]

I chose these words from Shakespeare not only because of their connection with the slavery induced sea suicide I refer to, but also because there is an inverted message in those words. I am reminding both myself and the audience that if we do not take care, it is not that bones will become coral, but that coral will become our bones. Not that eyes will become pearls but that pearls will become eyes. That we will suffer a 'sea change', we will *see* global change in which the poorest of the earth, the former slaves and the colonized, from Niger to New Orleans, will be the ones to suffer.[12]

Not only did the voices of all these dead writers become part of the text, but the process of using their books informed the plot. The character Moj emerged as a house slave, and the illegitimate daughter of her master, who has a secret passion for reading. Like most slaves, Moj is forbidden to read and write. Whilst she is supposed to be cleaning the bookshelves, she reads from them instead, taking in her master's/father's great library collection containing works by Marx, Darwin, Melville and others. Moj's reading is not her only secret. We discover Moj was taught to read by a field slave named May, who is her friend, her teacher and also her lover. When the master discovers a love poem, which Moj has written to May, he makes sure their punishment 'fits' the 'crime'. He whips Moj's writing hand and because May has read the poem, he whips her eyes. The result of May's horrendous head injuries is death. The horrific murder of Moj's lover moves her to escape, dressed like her father.

Like Ellen, Moj escapes to Boston and then sails to England. In a pub in Deptford, South London, Moj meets a sailor called William Black, based on the true-life William Brown who, according to the Annual register of 1815, was an African-Scottish sailor who served in the British Royal Navy for 11 years until 'she' was discovered to be female and possibly the first African woman in the Royal Navy.[13] Encouraged by William and inspired by Herman Melville's *Moby Dick* (1851), Moj gets a job on board a whaling ship bound for Antarctica. When she gets there, just like Captain Scott's men did years later, she paints her face black. She then removes her male clothing revealing her female body and sings a song to her lost lover. Then, in the famous final words of Captain Oates she says, 'I am just going outside. I may be some time', and walks out into the snow. Like Oates, Moj becomes an Antarctic anti-hero.

In using this style of devising/writing and citing deceased authors in creating the play, I was struck by certain ethical questions. Was it right to use other people's words, lives, deaths in my text? How can I ask permission if the person is already dead? This is still a debate I am having. So far in my thinking, I believe that theatre is always a collaborative exercise, and has the potential to be so even in the writing of plays. British theatre is traditionally perceived as a hierarchy where the author is god, a solitary genius, and usually a white man and Shakespeare, of course, is the god of gods. But many of Shakespeare's ideas emerged effectively from 'collaborations' with dead writers before him, such as Ovid and Horace. Indeed, sections of plays credited to Shakespeare may well have been written by his own peers. For me, as long as people's work is properly referenced and we acknowledge our influences, why not quote each other and use each others voices in our creative work?[14]

I felt the presence of the dead with me whilst I was composing the text at my computer not only through using sentences from deceased authors in the play itself, but through writing my own words. I remember reading Alice Walker describing herself as a medium when she was wrote the novel *The Color Purple* (1982). The characters came through her, and onto the page, like spirits. When sitting at my computer about to write the first words of the character Moj, I felt like I was sitting at a ouija board, my fingers hovering tentatively over the keyboard, then came her voice, first mouthed, then voiced and then typed:

> Maaaaassssa! Oh papa! So you weren't always a paaaastor.
> Dandy!
> Daaaaaaaddy!
> Look at you, looking at me, looking like you!
> In yours ruffs and cuffs velvet cloak top hat and cane like puss-y in boots,
> With the ladies and the liquor, and the tobaccaaah!
> Maybe you were a dancer–'de rum and de bum and de baccy!' *(song / dance)* before you were a defiler!
> *(Beat)*
> It's only ONE book you read NOW.[15]

It is at this moment in the script that Moj has found a photograph of her father/her master, which has been hidden away in one of the books she is secretly reading. This is the first time Moj realizes she looks quite like her father. A fact which will become important later in the story. The actual photograph is of me, dressed as Ellen Craft, shot by Del LaGrace Volcano on Antarctica, which is projected for the audience.

When I am performing and I look at the photograph of myself dressed like Ellen, it is deeply humbling. Many people have noted that I bear a resemblance to Ellen Craft herself. I feel her spirit with me on that stage. I look into my own eyes which look like Ellen's eyes, Moj looks into her father's eyes, which are her master's eyes, and the audience look at us all.

After my devising/writing process, it was time to move into the rehearsal room with the artistic collaborators I had invited to form the Antarctic Collective. *Moj of the Antarctic* was the first spoken-word play our Sheron Wray had directed as she is a dancer, teacher and choreographer – which is exactly why I wanted to work with her.

Figure 7.2 Mojisola Adebayo shot on location on Antarctica. Image © Del LaGrace Volcano, 2005

Everything is movement. Thought is movement. Dance is movement. Text is movement. When a baby is born she does not separate her thoughts, and gurgles and screams from her arms, and legs and head. We are fully integrated creatures. We have been taught in the West in particular to separate these functions out; to consider writing as a task of the mind and dance as a task of the body. This Cartesian dualism, like the concept of the binary, favours mind over body, male over female, Europe over Africa. We want to make work where text and body are one. Sheron watched how I spoke the text in the rehearsal room and extended the movement arising from the words into something approaching dance. The live production of *Moj of the Antarctic* therefore is a finely detailed total score in movement and words as well as original music by Juwon Ogungbe, design by Rajha Shakiry, lighting by Crin Claxton and visuals by Del, edited by Sue Giovanni.

This word and image text is, then, performed before an audience who, following the African aesthetic we seek and influenced by Augusto Boal, we encourage to interact with the show and help move the action on. The performance is natural and supernatural. It breathes with the living audience, and is inspired by the spirits of the dead, those to whom this play is dedicated. Ellen Craft, her ancestors and descendants and the Antarctic Collective, thank you for making my autobiography, this is for you.

Notes and references

1. Mojisola Adebayo, *Moj of the Antarctic: An African Odyssey*, in *Hidden Gems*, ed. Deirdre Osbourne, London: Oberon, 2008.
2. William Craft, *Running a Thousand Miles for Freedom: The Escape of William and Ellen Craft from Slavery*. Louisiana: Louisiana State University Press, 1999.
3. Homi Bhabha, 'Of Mimicry and Man: The Ambivalence of Colonial Discourse', in *The Location of Culture*. London: Routledge, 1998, pp. 85–92.
4. Stuart Hall (ed), *Representation: Cultural Representation and Signifying Practices*. London: Sage, 1997.
5. Alex Haley, *Roots: The Saga of an American Family*, New York: Doubleday, 1976.
6. Kwei-Armah <http://news.bbc.co.uk/1/hi/magazine/6480995.stm>, accessed 23 March 2007.
7. Josef Szwarc, *Faces of Racism*, London: Amnesty International UK, 2001, p. 39.
8. Max Jones, *The Last Great Quest: Captain Scott's Antarctic Sacrifice*, Oxford: Oxford University Press, 2003.
9. Adebayo, 2008: 156, paraphrased from Karl Marx and Friedrich Engels, *The Communist Manifesto*, London: Orion, 1996 [1848], p. 11.
10. Adebayo, 2008: 166.
11. William Shakespeare, *The Tempest*, London: Penguin, 2001 [c.1611]), p. 40.
12. The irony is that in making and touring *Moj of the Antarctic* I have deepened my carbon footprint and caused ecological damage. The challenge is to make sure these footprints lead toward consciousness raising on climate change. See George Monbiot, *Heat: How We Can Stop the Planet Burning*, London: Penguin, 2007.
13. See: <www.nationalarchives.gov.uk/pathways/blackhistory/work_community/docs/service_record.htm>, accessed 13 February 2009.
14. In that spirit let me 'big up' just some of the theatre artists who influenced *Moj of the Antarctic*: Ntozake Shange, Susan Lori-Parks, Debbie Tucker-Green, Percy Mtwa, Mbongeni Ngema, Barney Simon, John Kani, Winston Ntshona, Athol Fugard, George Bwanika Seremba, Samuel Beckett, Amani Naphtali, Patrice Naiambana, Lemn Sissay, John Millington Synge, Dario Fo, John Martin, Jacques LeCoq, Bertolt Brecht, Augusto Boal, Denise Wong and all in Black Mime Theatre, Emilyn Claid, Phillip Zarrilli, DV8, Wole Soyinka and the performers/makers of *Ubu and the Truth Commission*.
15. Adebayo, 2008: 157–8.

8

The Physical Journal: The Living Body that Writes and Rewrites Itself

Olu Taiwo

In this chapter, I intend to advance a way of conceptualizing living bodies from the perspective of a practitioner. The proposed concepts underpinned by my embodied memories are from lived experiences informed by movements derived from my body in its process of change. These memories neurologically construct a kind of virtual body, which is holographically projected as spatialized memory, helping my living body to write and rewrite itself. To articulate my thoughts to the reader, the reader will be introduced to new terms of reference which are part of a wider conceptual structure I call 'the return beat' (Taiwo, 1998: 157). This concept is defined as a cyclical experience of rhythm whose intervals feel like it magnetically returns to the centre of someone's physical journal.

Transcultural practice: a construction

From movements as an embryo, to the motion of our last breath, there is left a movement journal, a lived duration, a series of events imprinted in temporal spaces recorded in the time-lines of our ephemeral and subjective universes. Physical journals exists as sets of holographic memories that aid in manifesting our lived bodies, existing as a series of behaviours and movement pathways that help structure our physicality. They expose the complex nature of our identities in temporal space, starting with both our genetic inheritance and what has been nurtured by wider society. Our living bodies are created and structured as a result of drawing on embodied information from the knowledge and memory bases situated in our physical journals. Broadly speaking, the concept of the physical journal can be seen as the formation of a virtual body-mind that is neurologically constructed, similar to the formative structures that underpin a phantom limb; an awareness of which aids the reflective practitioner with embodied experiences in conjunction with personal practice.

The media for a written journal comprises of consciousness, intertextual ideas, a language, a word processors and paper, among other things. For the physical journal the media starts with effort as the writing tool inside a living body, and includes embodied knowledge and memory, an 'in-here-ness' and the nature of its relationship with the 'out-there-ness'. It incorporates learnt movements, shades of effort quality with postural expression. The techniques that I have written and embodied within my physical journal, which include; T'ai Chi Ch'aun, Capoeira, West African dance forms, Body Popping, Breakdance and Contemporary Release, is an example of a physical journal constructed from transcultural practice that is in the active process of creating new cultural expressions.

Transcultural constructions, underpinned by all these dance forms, in terms of the shape, content and cultural organization, has helped to structure my performative identity. Fernando Ortiz's concept of 'transculturation' (Ortis, 1995: 102–3) defines the phenomenal process of merging and converging cultures resulting in a transcultural practice created from any number of contexts that negotiates or not, the differences between more than one 'culture'.

The metabolic processes in physical journals are constructed and governed by conscious and unconscious efforts. The production of muscle is underpinned by transcultural practices under the contemporary umbrella of Globalization; practices that are increasingly producing new knowledge and effort maps. Consequently, with the construction of new social spaces and practices, a large proportion of our needs are produced by the forces of global capitalism. These forces are not altruistically concerned with the formative motivations of metabolic effort and given the choice, we will choose what we are programmed to select through mass media, unless we are sufficiently critical of the information that we are continually having to process, which means struggling with a balance between our internal and external perspectives. Ultimately, we are what we consume and what we embody in-forms our metabolic functions, which in turn in-forms our efforts. Ajayi highlights this key concept from a Yoruba perspective regarding the practice of 'counter-opposing forces creating a tension that equalizes resulting in balance' (Ajayi's, 1998: 27), represented by the Yoruba idea of 'ÌWÒN ÒTÙN ÌWÒN ÒSÌ', which when translated can be seen as the philosophy and aesthetics of symmetrical balance. This idea extends to every aspect of the Yoruba's worldview, whether civil, religious or artistic, similar to the Chinese concept of Yin and Yang.

Let us look briefly at the nature of informational knowledge in relation to a performer's physical journal. Broadly speaking there are two kinds of performative knowledges which are co-dependent and interwoven; these are knowledge of the 'out-there-ness' and knowledge of the 'in-here-ness': information that is latent or dormant, meaning that the information is not necessarily hard-wired into our conscious system, but has to be discovered and rediscovered via learnt processes. External information from our out-there-ness is gained empirically though experimentation, through our senses, by being in touch with what is observed. Internal information from our in-here-ness, is gained by intuitively investigating latent behaviours inherited from the past, including some latent and innate information stored in our DNA: some successful and not so successful perceptions and behaviours will be present as potentials. The excavated knowledge of both external and internal information is gained through experience. However, this distinction is arbitrary as the relationship between the 'out-there' and 'in-here' is more complex and interwoven.

There are differences between the term a 'living body' and a 'physical journal' in that the former refers to observations from a detached position useful for biological study, while the other suggests a personalized history that contextualizes our multiple and existential identities useful for expression. A 'living body' being autopoietic, meaning self-producing, has the ability to interculturally regenerate, interact and exchange information with itself, the environment and other organisms. In the West this raises questions about the 'conscious experience' of our constructed physical journals and how we articulate and share these encounters, bringing us to the concept of intersubjectivity.

Intersubjectivity and liminality

If we agree that each individual is a subject that contains conscious subjective experiences, then intersubjectivity can be seen as the combined corporeal state where two or more people are engaged in sharing information on many conscious and subconscious levels. When we integrate diverse actions originating from different subjects that include an account of persons, objects and areas of study, we have to include definitions of the various concepts concerning what a subject is and how particular subjects become joined due to their interactions. It can be argued that if the state of their being joined is intersubjective, then the holistic parameter of their conjoined activity is the state of intersubjectivity.

My main internal agitation is to redefine and guard metabolic effort against global forces in the twenty-first century: But more than this, it is a personal call, to emotionally reclaim 'ethnic' and 'indigenous' heritages within a transcultural framework (Harvey, 2000: 12). The seduction of technology that feeds our natural tendency toward an easier life makes redefinition even more critical, as substantial physical effort is not a prerequisite for a comfortable life within a contemporary Westernized household. New shades of effort in liminality are being made visible and conceivable with new technology, which makes performative actions and wise choices all the more valuable.

The nature of the liminal can be characterized as being associated with, or situated at, a sensory threshold. Susan Broadhurst adds to Turner's definition of the liminal as 'a fertile nothingness' or a 'storehouse of possibilities', to include a 'greater emphasis on the corporeal, technological and chthonic' (Broadhurst, 1999: 12). This a good place to start when attempting to articulate the dimensions and concepts surrounding my work, which performatively exists somewhere between 'form' and the 'formless', 'cultural juxtapositions' and 'hybridization', the 'metabolic' and the 'digital', 'individual' and 'social'. It lies at the juncture where the transformation of individual and group consciousness takes place. Critical theorists generally speak from the point of view of a participant observer, questioning the output and paradigmatic assumption of the practitioner in the field of contemporary performance. This is a valuable viewpoint, but it cannot and does not purport to theorize the journey of the practitioner from a performer's point of view. An important voice in this regard is Laban's in his work on choreological studies, which theorizes the performer's movements in temporal space from the dancer's perspective. When the body starts to move, the body's kinesphere is in an active relationship to other kinespheres and/or other objects in temporal space. The nature of these movement relationships, both their positions and orientations, prompt questions like: How does the physical journal orient itself? When we are moving, where is our conscious front? With gravity as a conscious reference point, how do we assess what is up and what is down?

The human genome: an owner manual

Bits of information from our DNA are responsible for the construction of psychological behaviour as well as our individual physiological properties. This amazing fact means that in the distinct field of shape and function, for example, of a blood cell and a liver cell, a complete library of instructions of how to make an entire human being is contained in both distinct structures separately. The different parts of the cell's physical journal are used (Kaku, 1998).

The Human Genome project has purported to reveal the coded information that lies at the heart of our DNA. When this information is fully understood, it is claimed, we will have an instruction manual that could help us understand how human beings are constructed physically and psychologically. Michio Kaku highlights this point by saying the decoded human genome will give:

> us an 'owner's manual' for a human being. This will set the stage for twenty-first century science and medicine. Instead of watching the dance of life, the biomolecular revolution will ultimately give us the nearly god-like ability to manipulate life almost a will ... [*He goes on to say that we will move*] ... from passive bystanders to active choreographers of nature.
>
> (Kaku, 1998: 9)

He is, of course, referring to the science of manipulation rather than the expressions and understandings of the 'Self'. It is an interesting fact that consciousness is at the frontier of science, which is linked to David Chalmers's, the Hard Problem (Papineau and Senna, 2000: 19). This raises the question of how conscious human beings can observe and analyse consciousness. Or more specifically, what is the nature of conscious experience? I do not intend to offer an answer to this hard problem here, but as a practitioner, the correlation between the above questions and the methods of the reflective practitioner produces an interesting link between science and art. The research into DNA so far suggests that we share most of our structural information with all the animals and plant life on earth. The information stored in us leads me to one of my propositions, which is that our bodies carry a large quantity of instinctual behaviours with evolutionary ancestries that lay dormant in our physical journal until triggered or researched. Releasing this dormant information requires a kind of personal archaeology, as well as the cultivation of various techniques which, when unlocked, increase new degrees of freedom producing new instinctual behaviours. However, there is a darker side to research into the human genome, where the environmental and ethical concerns are raised with projects like human cloning, and gene therapy to enhance performance.

Efforts, emotions and a perceptual hypothesis

When we discuss effort, we need to include in our discussions its connections with consciousness. Before any physical action, there is an 'inner stirring' in consciousness or an 'inner attitude' that creates the context for 'effort' leading to different effort qualities. Laban states that:

> The components making up the different effort qualities result from an inner attitude (conscious or unconscious) towards the factors of movement, Weight, Space, Time and Flow.
>
> (Laban, 1971: 13)

An effort quality, then, is not an abstracted expression operating out of context. For an effort quality to exist, an effort needs to be contextualized by answering a number of key questions. What are the reasons for the effort in question? Whose efforts are we referring to? Where does the effort come from? Why perform the effort at all? When is the effort being performed? These are only some of the questions we can ask when referring to the abstract concept of effort. The questions above, and some others not mentioned here, can be directed toward two main domains of action:

1. Towards goal oriented actions, that is, sporting, political and commercial endeavours.
2. Actions directed towards expressive goals, that is, artistic, religious and cultural events.

Without deconstructing these domains, we can agree on one fact, that a movement, as a result of some embodied effort, requires the satisfaction of an inner need or emotion to create the desire for the production of an appropriate effort quality. Embodied conscious effort is predicated by the existence of an organism with reflective consciousness, an experience of interiority with a physical journal. Therefore effort is embedded and embodied within a context for it to exist at all. Working with this assumption, we can say, as proposed by Laban, that the common instigator for our numerous endeavours attributed to our living bodies, accompanied by their physical journals, is 'effort'; embodied conscious effort (Laban, 1971: 13).

There is increased recognition that not only does the 'direct conscious experience', or Chalmers's 'qualia', play a central role in scientific research as a subject, it also plays a central role in the construction of feelings exemplified by neurologist Professor Antonio Damasio's work on emotion. According to Damasio, emotions precede feelings in the living body. Emotions are in direct response to inner motivations and drives, he writes with reference to the Dutch philosopher Spinoza:

> ... a number of drives and motivations. Major examples include hunger, thirst, curiosity and exploration, play and sex. Spinoza lumped them together under a very apt word, *appetites*, and with great refinement used another word, *desires*, for the situation in which conscious individuals become cognizant of those appetites.
>
> (Damasio, 2003: 34)

The subtle difference made between 'appetites' and 'desires' clarifies Damasio's precedence of emotions over feeling. 'Appetites' refer to the emotional behaviour of an individual engaged with inner motivational drives that have the potential of being public. 'Desires' on the other hand refers to an individual's private conscious experience of feeling; their becoming aware of a particular appetite, with the choice to act or not to act in order to satisfy the urge.

Through an examination of my effort, consciousness and emotion regarding artistic practice and teaching, I have concluded that what my consciousness observes, in the act of perception, is the data I collect from external clues via my multiple sensors. This cross-modal neural system transduces the information received, and decodes it to create a perceptual hypothesis. I then project this hypothesis into the judged location in temporal space (Velmen, 1998: 52). What I eventually 'see' is my construction of the out-there-ness, a construction that partly relies on the embodied information within my physical journal. Simultaneously, our senses are being continually bombarded with a massive amount of information at each moment. Much of this information is largely unknown to us due to the limitations of our sensors and the spotlight of attention which, being selective, does not necessary notice the peripheral information our sensors have scanned. However, in making sense of this cross-modal bombardment, Dr Gemma Calvert proposed some insights into the cross-modal or multi-sensory processes within our brains. Speaking on Radio Four's *In Our Time* (28 April 2005), she suggested that there are neurons in the brain that have learned over the course of evolution to detect the 'co-occurrence' and 'co-incidence' of spatio-temporal events. This is achieved by the self interfacing with multi-sensory information through the liminal flux in the out-thereness; registering both 'proxemics' and 'positions' concerning events in temporal space.

The physical journal and the haptic system

The living body's sensorium is conceptual, divided into eight senses that we use to interface with our out-there-ness. These include the Visual, Auditory, Gustatory, Olfactory, Cutaneous, Kinesthetic, Vestibular and Organic. There may be others, but for the sake of this analysis, I want to focus on three specific senses that Marilyn Rose McGee in her website refers to as being part of the haptic system. The haptic system relates to the total corporeal apparatus associated with touch. It is the sensory, motional and intellectual mechanism of a symbiotic relationship between the body, brain and mind. Without this haptic system, we would have no way of embodying or memorizing physical knowledge. The particular nature of this knowledge contains information regarding how our living body knows where different body parts are with respect to each other and how we orient, protect and balance ourselves in the out-there-ness. From the point of view of haptics, the informational clues received from the out-there-ness that is perceived and projected by the physical journal, is immediate, existing in 'now time'. One could argue that all the senses exist in now time, in that all phenomena perceived both internally and externally are in the moment. According to Rose McGee, the three senses out of the eight that make up the haptic system are:

- **The cutaneous sense:** Our skin's sensitivity to temperature, pressure and pain
- **The kinesthetic sense:** Our sense of moving our muscles, joints and tendons. Our awareness of the position of *our* limbs and the ability to perceive their active or passive movement
- **The vestibular sense:** Our awareness of balance, equilibrium and gravity. This system of fluids in the inner eye receives sensory signals from rotations in the field of gravity and accelerations and deceleration in temporal space.

(Rose McGee, 4 May 2005)

The combination of the cutaneous and kinesthetic senses synthesizes skin sensitivity and movement awareness to produce our perception of touch. This is an essential perception in the art of physical contact, whether in the martial arts or contact improvisation in dance. The other combination that Rose McGee considers is between the vestibular and the kinesthetic, which synthesizes the awareness of gravity and balance with movement to produce our proprioceptive perception. Without a proprioceptive sense, when we are blindfolded, we would not know where our limbs have moved to (ibid., Haptics).

From the above deductions, I have concluded there are three distinct areas of the physical journal that we can use to reflect on our cognitive faculties in movement. These cognitive faculties use this haptic network to regulate the production and presentation of energy, effort and movement in temporal space. When I say movement plays a major part in the formation of thought, thinking and feeling, I am shifting beyond the idea of movement as a performative result of these cognitive faculties, to movement as an essential building block for the creation and appearance of thought, thinking and feeling.

The three distinct areas of the physical journal in temporal space are illustrated in Figure 8.1.

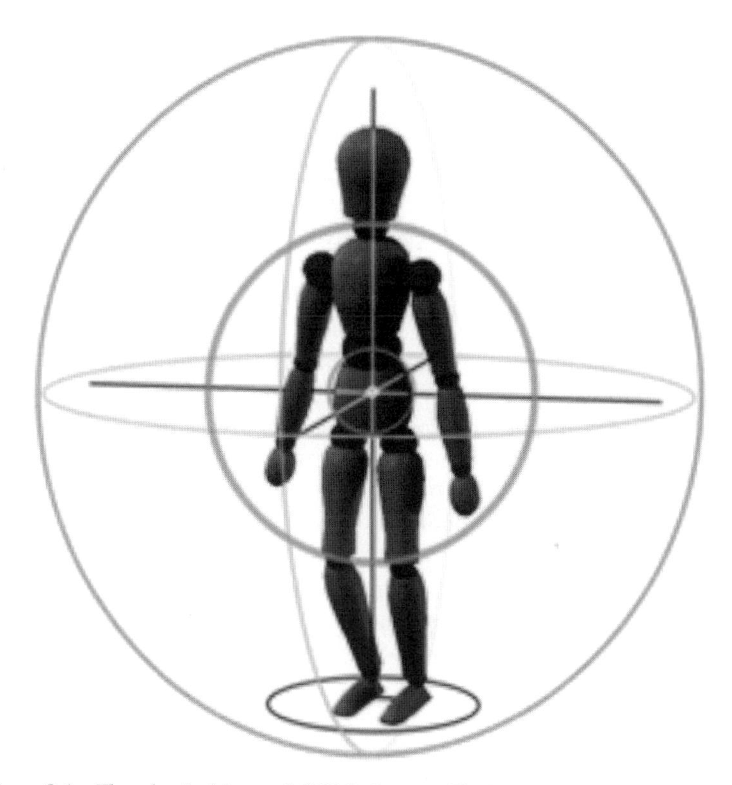

Figure 8.1 The physical journal (2006). Image: Olu Taiwo

1. Stable core (*central area*).
2. Temporal space in the body (*internal area*).
3. The body in temporal space (*external area*).

The stable core (*central area*)

This defines the performer's core stability and is when the performer reflects on the mental and physical forces of effort acting within the human frame. Core stability relates to rhythmic balance and anatomical alignment as a way of optimizing the effort to action ratio. Around the body's centre is a small sphere whose edges take in the abdomen and hips. Within this sphere are reference points radiating from the body's centre, used in the maintenance of core stability in motion. This requires an organizational intelligence, which requires the performer to reflect on their neuromuscular skill when they are articulating and sequencing movement. This is a kind of grammar for proficient movement.

Temporal space in the body (*internal area*)

This is when the performer reflects on the internal forces of effort acting within the kinesphere. This highlights the synthesis within the haptic system producing our proprioceptive perception within the sphere of movement around the performer's living body (Laban, 1966: 8). In other words, reflecting on the pathways through which the performer's limbs move with their zones of expression. What we are talking about is the internal area dominated by proprioceptive memory. This is the ability to remember a sequence of movements; for example, in Capoeira there are a number of movement sequences that have to be embodied. These sequences have a coded intelligence based on the strategy of a martial dance. Once the physical journal has embodied and understood the logic behind these sequences, the ritual dance with a partner in the 'roda' (dance circle) can comprehended. This empowers the performer 'to conceive' in terms of movement by mentally reconstructing muscular effort, motion qualities and positions post-choreographically in space. Reflecting on these skills opens up degrees of freedom for expression and action based on core stability. The kinesphere marks out the boundary of the practitioner's mobile studio, which becomes a studio in any situation, only through the embodied practice of being mindfully reflective.

The body in temporal space (*external area*)

This is when the performer reflects on the forces of effort acting externally within a shared general space; the stage, the court, the field, work and play. This area is the symbiotic nexus between the complete haptic sensorium and our out-there-ness. This is where with core stability the physical journal regulates effort between the torso, limbs and the kinesphere through the performer's proprioceptive memory. With this embodied facility of regulation, the practitioner acts as one intuitive body to move and react within a shared environment. What is essential is the performer's capacity to reflect on their proprioceptive awareness in the moment. The whole, then, gives moment-to-moment feedback to the performer, who is practising being 'mind-ful' and 'body-ful'. This includes postural alignment, change in the body's equilibrium, changing position in relation to other people and objects in space.

The three areas are interwoven into an integrated state that underpins the parameters of the practitioner's mobile studio practice. The spherical edge of the mobile studio I refer to as a perceptual membrane corresponding to and expanding Laban's concept of a kinesphere. The use of the word membrane is important here, as it evokes images of the biological interfaces that circumscribe the cell's crystalline membrane. However, in this case the membrane is defined as being 'virtual' and is neurologically projected.

Mobile studio practice

The notion of the mobile studio cunningly shifts the conceptual focus of the studio from the body in temporal space, that is, the dance studio, to the temporal space in the body, the kinesphere. More importantly, the mobile studio provides the potential for a disciplined personal zone, to develop our physical journals by revisiting our effort drives and movement pathways using the three distinct areas. The aim is to re-evaluate our inner habitual attitudes or engrams towards our choices and decisions in effort, to allow new motivations to develop and form new relationships between the body and a changing environment. We do this so that we can trigger and release dormant physical texts in different environments. This requires particular patterns of behaviour that enable the practitioner to research develop and reflect on their cognitive faculties in movement.

The combined relationship between the mobile studio practice and the physical journal is further layered by a desire to understand the nature of movement phrases by looking at 'trace forms' from three points of view. Laban describes these different views as being from the perspective of the 'emotional dreamers', the 'scheming mechanics' and the 'biological innocents' and adding that these three 'layers were rooted in our inner life' (Laban, 1966: 7). For my part, the advent of new technology has expanded our ability to review Laban's orthogonal perspective of analysing movement phrases. We can see the corporeal mobile studio as the ground plan, with the imaginative studio and the digital studio as interchanging with the side and end elevations. This, of course, is only a metaphor. All three perspectives in temporal spaces have their own specific techniques of reflecting choreographic and post-choreographic activity to create movements that express emotions, ideas and aesthetics.

1. The imaginative studio.
2. The digital studio.
3. The corporeal mobile studio.

The imaginative studio is a particular process that most sports practitioners and physical performers employ, whether consciously or unconsciously, to rehearse and practise a technique they have embodied in the physical journals before a performance. It is also a place where we access the internal world of our imaginations relating to Laban's concept of 'emotional dreamers'.

The digital studio is when we record and translate the motion information from our physical journals, by using motion caption or traditional video into digital data. When objects or events are reduced into digital data, they becomes subject to infinite changes, giving a practitioner an infinite amount of analytical control. This 'objective distance' gives us the possibility of recognizing and designing patterns that have intuitively emanated from our imaginations. This relates to Laban's concept of 'Scheming Mechanics'.

The Corporeal studio – is about physically and psychologically developing effective movement engrams suited to the living body's context. An engram is an embodied memory located in the complex relationship between the nerves and muscles usually created by repetition and sometimes through shock. Primarily though, the relationship between the nerves and muscle are about the sensations felt by the practitioner, which further informs the embodiment of new habits, relating to Laban's concept of 'Biological Innocents'.

In conclusion, the concept of a physical journal is defined as a neurologically constructed body-mind that is virtually projected similar to the idea behind the structure of a phantom limb. This body-mind is the individual's physical journal; layers of temporal space folded inside their perceptual membrane.

An important point about the 'mobile' in mobile studio practice is that as practitioners/students/lecturers, we should recognize the importance of developing an appropriate and serious practice to keep our physical journal at a state of readiness suited to our needs: including how and when we shift into the zones of practice as an extra daily activity. This is because the state of our physical journals is always in flux, it is a dynamic process where our efforts struggle between the influences from our 'in-here-ness' and clues from our 'out-there-ness'. Engaging in flux, the physical journal has the ability to write and rewrite itself and therefore change aspects of the living body. To Laban, observation of flux within the flow of bodily action is the fundamental factor when we are internally and externally perceiving, performing and observing movement in temporal space. Flux here is seen as the abstract property of continuance describing the progressional nature of how movement content changes and flows.

The flux of movements, resulting from 'effort' and 'experience' when balancing the layers in the physical journal, can draw attention to our lived experience of change as we live, perceive, conceive, produce and perform. So in referring to the nature of flux, I am drawing specific attention to the phased and balanced link between the continuous flows of change within the physical journal, mobile studio practice and the textures of and in temporal space.

References and suggested reading

Books

Ajayi's. (1998). *Yoruba Ritual.* Trenton, NJ, and Asmara, Eritrea: Africa World Press.
Barba, Eugenio and Nicola Savarese. (1991). *A Dictionary of Theatre Anthropology: The Secret Life of a Performer.* London: Routledge.
Barthes, Roland. (2000). *Mythologies.* London: Vintage.
Broadhurst, Susan. (1999). *Liminal Acts: A Critical Overview of Contemporary Performance Theory.* Oxford: Cassell.
Da Liu. (1986). *T'ai Chi Ch'uan and Meditation.* London: Arkana.
Damasio, Antonio R. (2003). *Looking for Spinoza.* London: Vintage.
Deleuze, G. and F. Guattari. (1994). *What is Philosophy.* Trans. H. Tomlinson and G. Burchell. New York: Columbia University Press.
De Quincy, Christian. (2002). *Radical Nature.* Motpelier, VT: Invisible City Press.
Drewal, Margaret Thompson. (1992). *Yoruba Ritual.* Indianapolis and Bloomington: Indiana University Press.
Edelman, G. M. and Guilio Tononi. (2001). *Consciousness: How Matter Becomes Imagination.* Harmondsworth: Penguin.
Gilroy, Paul. (1993). *The Black Atlantic.* London: Verso.
Harvey, Graham (ed.). (2000). *Indigenous Religions*, Oxford: Cassell.
Kaku, Michio. (1998). *Visions.* Oxford: Oxford University Press.
Laban, Rudolf. (1966). *Choreutics.* London: Macdonald.
——. (1971). *The Mastery of Movement.* London: Macdonald.

Lefebvre, Henri. (1994). *The Production of Space*. Oxford: Blackwell.
Lewis, J. Lowell. (1992). *The Ring of Liberation: Deceptive Discourse in Brazilian Capoeira*. Chicago: University Of Chicago Press.
Murphy, T. (1998). 'Quantum Ontology: A Virtual Mechanics of Becoming', in *Deleuze and Guattari: New Mappings in Politics, Philosophy and Culture*. Ed. D. E. Kaufman and K Heller. Minneapolis: University of Minnesota Press.
Ortis, Fernando. (1995). *Cuban Counterpoint: Tobacco and Sugar*. Trans. Harriet de Onis.Durham, NC: Duke University Press.
Papineau, David and Howard Senna. (2000). *Introducing Consciousness*. Cambridge: Icon.
Preston Dunlop, V. and A. Sanchez-Colberg. (2002). *Dance and the Performative: A Choreological Perspective – Laban and Beyond*. London: Verve Publishing.
Ray, Christopher. (1992). *Time, Space and Philosophy*. London and New York: Routledge.
Sanchez-Colberg, Ana. (1996). 'Altered States and Subliminal Spaces: Charting the Road towards a Physical Theatre', *Performance Research*, 1.2 (Summer): 40–56.
Soyinka, Wole. (1993). *Art Dialogue and Outrage*. London: Methuen.
Taiwo, O. (2000). (1998). 'The Return Beat', in *The Virtual Embodied*. Ed. J. Wood. London: Routledge.
——. 'Music, Art and Movement among the Yoruba', in *Indigenous Religions*. Ed. G. Harvey. Oxford: Cassell.
Van Nieuwenhuijze, D. O. (1998). *The Simplicity of Complexity*, Conference Paper at Fourteenth World Conference of Sociology, Montreal.
Velmen, Max. (1998). 'Physical, psychologies and virtual realities', in *The Virtual Embodied*. Ed. J. Wood. London: Routledge.
Wenger, Susanne and Gert Chesi. (1983). *A Life with the Gods: In their Yoruba Homeland*. Woergl: Perlinger.
Whitehead, A. N. (1975 [1929]). *Science and the Modern World*. Glasgow: Fontana.
Wile, Douglas. (compiler and translator). (1983). *Yang Family Secret Transmissions*. New York: Sweet Ch'i Press.
Wood, John. (1998). 'Reinventing the Present', in *The Virtual Embodied*. Ed. J. Wood. London: Routledge.
Yang Jwing-Ming. (1987). *Advanced Yang Style Tai Chi Chuan*, Volume 1.United States: Yang's Martial Arts Association.

Websites

Derrida Jacques. (1982) Excerpt from *Différance*, in *Margins of Philosophy*. Trans. Alan Bass. Chicago: University of Chicago Press, pp. 3–27; footnotes are not reproduced. <http://evans-experientialism.freewebspace.com/derrida5.htm>, accessed 24 June 2009
Maturana, Humberto. Autopoiesis. <http://www.oikos.org/mariotti.htm>; <http://courses.nus.edu.sg/course/elljwp/derriduction3.htm#alterity>, accessed 20 June 2009.
Phillips, Maggi. *The Embodied Thesis*. AARE Mini-conference 2003
Conference Papers.<http://www.aare.edu.au/conf03nc/ph03022z.pdf>, accessed 20 June 2009.
Rose McGee, Marilyn. (04, 05, 2005), Haptics. <http://www.dcs.gla.ac.uk/~mcgeemr/work.html>, accessed 20 June 2009.

Part 3

Performing the Body/Performing the Text … Writing the Body/Writing the Text

9

Socializing the Self: Autoethnographical Performance and the Social Signature

John Freeman

The further we move towards control over the reproduction of images, events and experiences, the more enhanced, perhaps even desperate, our attempts at capturing the real have become. As the site in which the represented other of character is made manifest through the viscerally authentic performer, live work has always shown a tension between self and the shadows it throws; and no form is at the same time as shadowy and substantial as autobiography. Because the search for the defining features of autobiographical performance has become something of a subindustry in itself – a critical search to equate the paradoxical elements of facts and subjectivity, honesty and artifice – this chapter needs to come clean about what it will be attempting to add to the field, and also what it will not.

As an intrinsically borderline genre, located between experience and memory, reportage and invention and, to paraphrase Schechner, between cooked art and raw life (see Schechner, 1988: 191), the question of whether or not the events articulated in any given autobiographical text are necessarily true or false is not of particular importance in the following pages. The emphasis here is on discursive elements of representing aspects of a person's life through the autobiographical genre in ways that speak to wider communities, and that do so whilst acknowledging performance in its fullest sense; not as a means through which one can escape from the realities of life through sentimentalized entertainment, so much as a method through which we are encouraged to contemplate life imaginatively, deeply and with increased shifts in understanding. These shifts occur as soon as we acknowledge understanding as a 'series of perspective shifts' rather than as a 'once-and-for-all act' (Owens, 2007: 3). We know that whereas mainstream drama was at one time (and often still is) informed by critical perspectives born out of the literary canon, contemporary performance is now largely informed by a canon of post-structural theories. The performance focused on in this chapter is informed by a broad application of these theoretical interventions, each of which posits self-reality as no more than a composite of historically situated language constructs within which our individual or subjective perspectives are programmed to play their part. We witness this on a regular basis with autoperformers, to borrow Michael Kirby's term (1969: 155), those practitioners whose work conforms to post-structuralist ideals by presenting material that acknowledges and investigates the relationships between subjective perspectives and the social-historical manifestations of construction. So far so familiar, but this chapter's post-structuralist drift is tempered by the growing belief that a great many practitioners and academics have become so insulated from the real-world consequences of their work that they (we/I) are inclined to speak to each other in papers and presentations that are little more than love letters to colleagues and/or coded applications for promotion. This creates a cycle of what is often little more than the language of oppression, obscured by the babble of hypersubjectivity; or, as Noam Chomsky sees it, as hypersubjectivity as a form that seeks to deprive the working classes of the tools of emancipation, playing working people and vulnerable, excluded communities against each other through claims that all projects of enlightenment are redundant in so far as inequity is never anything more than one illusion in a world of simulacrum (see Chomsky, 1993).

Foucault's observations that 'truth' is an effect of power have created telling critiques on notions of objectivity, but where oppression is a fact of daily life placing inverted commas around the word 'truth' adds casual intellectual insult to savage social injury (Foucault, 1980), If a weakness with much work we see is that the events that we are presented with are less vital and somehow less real than our experiences of actual life, it follows that we are almost inevitably disengaged from that which we see, if for no more reason than that we have already lived through considerably more than most performance imagines on our behalf. It is in this absence that much seemingly progressive performance has left us with a space where intentionality used to flourish. We have developed a critical terminology that conceals this absence and in much of the peripheral micro of my own previous writing as well as in the more substantial macro of others' it is one that even serves to turn absence into a virtue. And yet this restless vocabulary obscures that which might still be the overriding function of performance making. This is to present some element of the struggle between intention and obstacle so effectively as to make us understand, in the truest sense of the word, those same struggles that we routinely fail to comprehend, no matter how relentless our daily engagement.

With powerful, effective work this experience is reversed. Those same situations and circumstances, tensions, intentions and motives which we spectators have been unable to make sense of through either imagination or experience are realized for us in and through performance. The most important questions, then, are not about how accurately a specific play reflects the look of life, or how heavily the blood flows from a performer's self-lacerations, nor even how zeitgeist-connecting we find the *mise-en-scène*. What matters is how much the work allows, supports and encourages its audience to understand and grapple with the meanings of life as we/they live it. This is a grand-sounding statement, and one that seems at odds with our twenty-first-century embrace of anything goes (anything perhaps except this). And yet writing the words does not feel like a reversal or return. If recent performance practices have started giving us much of the same, then at least that incessant sameness has brought with it a broad acceptance of work that is no longer limited by the view that all performance work other than that which stems from dramatically scripted representations of 'other' is essentially peripheral and reductive. The idea, rooted deep through autobiographical and confessional portrayal, that theatrical (re)presentation can be primarily concerned with the revelation and not the concealment of the performer's self – a very different type of 'other' – has opened us up to new forms that have left presumably indelible traces of discourse-privileging, ironic distancing, denial of closure and the mixing of high and low cultural forms, and these have become as much a part of our contemporary performance landscape as naturalism was one hundred years ago. The challenge now is to create a new ethically informed practice: one that approaches self as a social subject. In response to this challenge a case will be made that what matters most in autobiographical performance is not necessarily a work's record of an authentic historical past (assumedly the lived experience of the author/performer) so much as the ways in which it functions as a form that is fundamental to the construction of modern life.

The approach to performance adopted by John Leguizamo in his December 2003 Broadway Theatre performance of *Sexaholix ... A Love Story* will serve to illustrate this, and a reading of his work as it relates to autoethnography will inform the second part of this chapter. Best known for his high-profile film work, Leguizamo is not an immediately obvious practitioner to draw upon here. There is a form of logic behind the choice that goes beyond spectatorial and critical appreciation:

Leguizamo and *Sexaholic* function in many ways in this chapter as examples instead of paradigms. The distinction is based on the notion of the paradigm as a pattern, model or exemplar, as a practice that is used to define a discipline during a particular period of time. This is in opposition to an example, which is more commonly used to refer to a representative of a group or concept. It is important to stress that no such claims are being made for the work described in this chapter to stand as intrinsically paradigmatic. This is not a qualitative decision and it is certainly not intended as a caveat. The standard of the performance work undertaken by Leguizamo is not in question in this chapter. What matters is a cautious approach to the seeming embrace that comes with the territory of paradigms – a type of paralysis; an inability or unwillingness to see beyond the current modes of thinking, a closing down of the potential for radical differentiation that is not hierarchically bound.

With paradigms the new inevitably emerges, which replaces and supplants the old. This implies a value system that this chapter is at some pains to resist. Quality notwithstanding, the example cited here is not necessarily 'better' than any others: just in key ways different. The relative dearth of critical material regarding Leguizamo's live performance work is another means of nudging at the edges of accepted paradigms. Ultimately, regarding examples as paradigms suggests a dependence on judgement that goes beyond the editorial assessment of what material to include and what to omit. Added to this is the somewhat larger issue of the paradigmatic concept's appropriateness or inappropriateness for discussions of performance practice (see Kuhn, 1962).

With autobiographical work performers feel compelled to go public because they believe that their lives matter, that their stories deserve and even need to be told. In both content and form, autobiography is drawn from this moment of realization, liberation and externalization. Autoethnography differs in that whilst it can present an individual performer's perspective, it is one that draws on and connects to a collective understanding. It is a site for recognizing communities as well as individuals, for understanding self as innately relational. It is in this recognition that presenting one's life through performance is able to foster a sense of community that comes with sharing rather than seeing life experiences, and with responding with an increasingly empowering engagement instead of merely rewarding the work with what is at best empathetic applause. Autoethnography is thus related to autobiography and to ethnography.

Ethnography is most commonly known as a qualitative research method where researchers employ participant observation as a means of achieving understanding of a particular group's culture. The researcher is most usually a member of the group in question, albeit on a temporary membership, rather than adopting a normative and assumedly disinterested stance. Autoethnographical research stresses these group allegiances, leading to analytical reflections on one's own experiences located within an identified cultural grouping. Deborah Reed-Danahay presents autoethnography as a challenge to the objective observations of ethnography wrought through postmodern autobiography, within which any notion of a coherent, individualistic self has been called into question. For Reed-Danahay autoethnography;

> has a double sense – referring either to the ethnography of one's own group or to autobiographical writing that has ethnographic interest. Thus, either a self (auto) ethnography or an autobiographical (auto) ethnography can be signaled by 'autoethnography'.
>
> (Reed-Danahay, 2007: 412)

Autoethnography in the performance terms that apply to this chapter is about work that draws on personal experience, but which foregrounds this as a meaning-making endeavour that aims to engage the broader social issues of a given community. As has been previously noted, we are not talking here about performative events which sincerely testify to certain truths. On the contrary, we are considering staged works which through their claims to be accepted as truth-performatized, as narratives that ostensibly explore real experience, and which strive purposefully and with the help of certain recognizable strategies to stress the referential authenticity of the work, create quite distinct conditions for spectatorial responses. In doing so, it will be argued, notions of applied and community-driven drama are being fused with live art sensibilities in ways that are seeking to restore purpose to performance.

We have seen in recent years – and whisper it soft – hints of a post-postmodern collapse of the subversion-by-rote of much critical thinking (see Martin, 1990), of a merciful release from the need to place every potentially contentious or unstable word inside single inverted commas: a need that by my count would have seen seven examples from this chapter's opening paragraph framed in this way. Partial freedom from the Punctuation Police is one thing, another still is freedom from the divide that has existed between the regularly encountered extremes of the potentially patronizing and the painfully pretentious; of drama as a 'magic bullet' (Owens, 2007: 7) fired at a captive audience on one side of the fence and performance that withdraws all elements of theatrical interest and which seeks no justification other than its own ontology on the other. There is more (and no more) to this than wordplay: at the stubbed out cigarette end of postmodern we may well be, and yet it is this same postmodern performance theory that has deconstructed the status of the author in the process of signification to the extent that autobiography is seen as being nothing other than a trope, with no crucial reference whatsoever to reality. And this is a powerful legacy.

A particularly incisive and influential reading of autobiography has been made by Philippe Lejeune who, in his concept of the autobiographical pact stresses the importance of the author's signature alongside authorial intentions (see Lejeune, 1989). For Lejeune authors are of fundamental importance in that it is they who create the discursive aspects of an autobiographical pact with the reader. Certainly, we need to substitute spectator for reader and performer for author, but these are simple exchanges and they can be envisaged without damaging the integrity of Lejeune's ideas. One of the immediate consequences of the concept of the autobiographical pact is that it demands the participation of at least two parties; the performer/author and the reader/spectator. The concept implies negotiation, participation and exchange. It also implies a move from autobiography to autoethnography.

Authoritative performance scholars such as Deirdre Heddon (2008), Sidonie Smith and Julia Watson (2001) and Elaine Aston and Geraldine Harris (2008) have so comprehensively theorized confessional presentations and/or that which takes the self-as-subject that we know well the etymology of autobiography as 'self-life-writing', from autos, bios, graphe; autoethnography differs fundamentally in the switching of bios (life) for ethnos, meaning people who share a distinctive cultural identity, and it is on this axis of shift that we are moving away from our recent embrace of catharsis for the artist as an *ipso facto* indicator of worth.

Postmodern performance built (if we can use the past tense already) on certain narrative forms for realization, at the same time as it toyed with a divergence from traditional forms in its drive for resistance. The boundary between these forms and other types of theatrical performance became somewhat diffused, creating a liminal space, a type of void into which increasingly formulaic performance was cast. It is within just this space that autoethnographical performance is developing as the form of post-postmodernism mentioned above: as postmodernism with a social conscience. Autoethnographical performance is a form of self-narrative that locates the self within social contexts. In this, it is both method and text, a form that is increasingly utilized by performers who place the story of their lives within recognition of the social contexts in which they occur. In autoethnographical work the importance of writing becomes a central concern, circling and even undermining the text as it creates it, invoking Lejeune's awareness of pact with Deleuze and Guattari's figurations of the rhizome and haecceity as ways to rewrite and re-right personal subjectivities within performance (Deleuze and Guattari, 1987). The performative qualities of being relational, bound always by the criteria of positionality, are located also within culturally bound, social and historical circumstances, with the relationship of performers to the people to whom the text is addressed and presented. It is in this concentration on the socialized selves involved in autoethnographical performance writing that a new form of twenty-first-century functionalism emerges, heralding a shift towards art that acts not as an agent of self-awareness or social change so much as a reconsideration of terms such as 'empathy', 'engagement' and 'community'. Terms which, in recent years, have become increasingly marginalized in the field of progressive performance writing and making.

This takes us slightly away from the default cultural touchstones of Barthes, Derrida, Foucault et al. For many, the postmodernist embrace of the call that the author is dead shifted the focus away from the act of making to the activity of reception. This in its own turn prematurely foreclosed the question of agency for a number of groups. That which makes one type of sense from a white, male, heterosexual perspective is not necessarily common to all. As Donna Stanton notes, a consequence of gender-bound discursive situations in which feminist artists are engaged is a commensurate need to 'privilege and promote the female signature', aligned to a need to 'make it visible and prominent, or else endure and insure more of the phallocentric same' (1987: 16). In sections such as these Stanton is remarking on the historical and increasingly well-documented fact that written texts by women, no less than women's performance works, have remained invisible in the canons of literature and performance far more regularly than those created by (and often for) men. The aim to analyse representations of socially committed subjectivity does not mean that we hope to find anything even approximating to a genuine self in any given works of performance. This writer, for one (and I suspect like many of us in performance studies), would be painfully unqualified to distinguish the real from the representational other than through the sound-bite knowledge that comes with borrowed terminology and the glib (re)assurance of publication. The intention is to interpret in what ways particular, often localized experiences are represented in the public discourse of autoethnographical performance.

Where autobiographical performance demonstrates its innate historical and cultural situatedness through which identity, experience, memory and craft are endlessly negotiated and refashioned, much solo performance work downgrades those same structural inequities which first prompted the artist to articulate the self. These inequities of class, ethnicity, age, appearance, gender, sexuality, finance, education, access and so on are too often sacrificed to the twin cultures of confession and trauma that are beginning to infuse every aspect of contemporary life. Performances dealing with trauma have marked out their territory within the broad realm of victim art through their narration of instances of individual abuse. And there is clearly a legitimate place for work of this kind. Autoethnographical work, however, is better located to address representations of the gendered, classed and raced experience, prompting and engaging ideas of community, individuality and belonging, exploring and exposing the ways in which particular freedoms are granted and/or restricted to particular social groups. It is through the articulation and demonstration of these experiences of individual social mobility that the ways in which concealed privileges are deployed in order to enable or constrain are exposed to scrutiny, if not quite made ready for change.

A number of performative modes exist with the aim of showing the ways in which identity and self are social constructs and which seek to expose the act of performance as a constructed and value-driven phenomenon. In this way performers are able to critique certain constructs at the same time as they are contextualizing them within broader social-political frameworks. Autoethnographical performance is one such mode, one that reveals its material as autobiographical at the same time as it foregrounds the attempt to communicate matters of shared interest to spectators based on societal rather than primarily intellectual or aesthetic concerns.

Leguizamo's early performances *Mambo*, *Spic-o-Rama* and *Sexaholix ... A Love Story* self-consciously portray his life as historically and psychologically representative of all immigrants to the United States, at the same time as they draw on his own distinctively Colombian heritage. Leguizamo resists the Western characterization of Spanish-speaking Americans, choosing instead to satirize racial and ethnic stereotypes in American culture. His most recent main-stage work, *Sexaholix*, is at once an example of autobiographical catharsis through its reconstructing and revisiting of a life history at the same time as it is an autoethnographical study of representation, translation and authorial agency. Where autobiography looks to tell the unique life-experiences of an individual 'I', the same 'I' becomes plural through autoethnography. It does so inasmuch as the performance is made to stand for a community of people who share a common identity and representative experiences.

Works like these are the 'outlaw genres' identified by Caren Kaplan (1992) performances which compose strategically staged events for those artists who position themselves or who are themselves positioned beyond the parameters of normative theatrical empowerment. Guillermo Gómez-Peña and Coco Fusco are noted examples of performers whose work provides often exhilarating examples of dramatic self-representation that ask tough questions of autobiographical representation, authenticity, socio-political comprehension, linguistic and cultural boundaries and audience expectations. Fusco and Gomez-Peña's work is well-documented in books such as Gómez-Peña (1994, 2000, 2005) and Fusco (1995, 1999, 2001). What we see in the work of these three practitioners is a playing with the edges of the boundaries between some of the aesthetic considerations that come with the territory of 'academy art' and socially committed performance.

There is logic to this seeming split, and it is a logic of estrangement and empowerment that comes with being at once always in key ways outside of the professional world in which one moves. Not for nothing are outlaw genres so called. The term implies a shifting of the autobiography-by-rote of elitist postmodern performance towards a complex body of exchange between spectacle and spectator that explore issues of identity, self-representation and self-recognition. That this approach demands a different understanding of fitness for purpose is axiomatic. Leguizamo's performance allocates a series of idiosyncratic gestural codes to each of the many characters he presents. This physical shaping allows his spectators to keep a degree of pace with the frenzied comedic virtuosity of the work. It also generates the type of energetic verve not usually seen in the overtly self-regarding work most familiar to audiences for autobiographical art house performance. Leguizamo himself makes few, if any, grandiose claims for his work, preferring to keep things light: when pushed by spectators of early readings of his work at New York City's not-for-profit performance venue PS 122 to make his material 'darker and deeper', Leguizamo's response was to take his work further towards hilarity (Leguizamo, 2007: 190).

Fundamental to our reading of Leguizamo's work as autoethnographical (despite his own occasional claims against such loaded terms) is our knowledge that social roles and their attendant values are communicated and absorbed through a chain of symbolic actions that are central to that which might elsewhere be usefully termed 'the body politic' and which for the purposes of this chapter we can describe as 'the body social'. This is not intended to mark some crude distinction between society and politics so much as it is a way of noting the performing autoethnographical body as the locus where cultural groups are able to share in, preserve and perpetuate their values. Leguizamo may see *Sexaholix* as means of getting closer to the public revelation of an already semi-public self and little more (Leguizamo, 2007: 257), and yet his treatment of this same self through an epic storytelling mode is sophisticated and effective enough to allow the performer to review past experiences, as part of a complex construction of present identity that utilizes elements of contradiction and parody to tap into collective memory and common rites of passage. When Leguizamo describes his childhood experiences of the 'Fresh Air Fund', an organization 'that takes poor underprivileged inner city kids and sends them to a rich white family in the country for two weeks' (Leguizamo, 2008: 255) his Broadway audience is split between those who laugh at the delivery of the line and those who laugh differently at what the line delivers, seeing their own experiences played back to them.

Alan Read points to certain types of performance that take as their material 'the neglected and the undocumented' (1993: 2) and this is what *Sexaholix* provides. We can say that the communicative body, like the performative body, is exemplified by the act of engaged, open and democratic communication, by the creative act and event of autoethnography. Past experiences are reimagined in the told situation of their stories and this is done in ways that are intended to evoke empathy. Through his work, Leguizamo performs and also advocates a communicative ideal. His performative rhetoric is therefore open to spectators rather than being closed off or solipsistic. Through work that is endlessly contingent, flexible and rooted in a play of communicative forms, the one thing we are not witnessing here is a prioritizing of the idea of catharsis for, by and in honour of the artist.

On the contrary, we experience, according to Mary Brewer, the type of performance through which we witness 'one of the cultural frameworks within which the spectator may recognize his/her location and level of complicity in ... social conventions' (2005: xiii). The postmodern rejection of 'the idea that assertions about the natural or social world can be objectively (and hence transculturally) true or false' (Sokal, 2008: 269) falls into disrepute when one considers Bertrand Russell's claim that the 'concept of "truth" as something dependent upon facts largely outside human control has been one of the ways in which philosophy hitherto has inculcated the necessary element of humility. When this check upon pride is removed, a further step is taken on the road towards a certain kind of madness' (Russell, 1991: 202). Whilst an autoethnographical performance writer like Leguizamo might have an initial intent to explore aspects of self, this is also about the need for a place to share that voice.

In the context of performance, creating text as a form of self-knowledge is always more than writing the individual self, it is about voicing that which power has silenced. In working the material of a life through autoethnographical forms the individual's story is made to connect with those of others. The writer/performer is possessed of an individual voice, but this is made to connect to a collective experience and a collective voice, and it is at this point that a sense of community can be born through autoethnography. Shifting our focus from performative self-orientation as intrinsically private activities for public spaces, we are able to view autoethnography in the context of reigniting common values, of tapping into the values of communities through sharing stories that matter to more than the storyteller; that speak from the body of individualistic recall to the body of collective experience. If it is not quite true to say that until a life is shared through writing or performance it does not exist at all, we can at least say that without performance, one's life is not able to resonate in the broad realm of public consequence. If public consequence is linked to social commitment then performance needs to demonstrate its own commitment to society, and this involves more than playing to a society of scholars.

Ultimately, autoethnographical performance is as much about a socio-philosophical ideal as a direct echo of experience. We know after all that it is not necessary to have the exact same experience as someone in order to connect with their stories, just as we know that empathy is possible through both similarity and difference. What matters most is that performance based on reconstructions of writerly selves is harnessed to transformational identity politics in ways that pursue connections with audiences, rather than regarding spectators as being fortunate simply to witness our work.

References and suggested reading

Aston, E and G. Harris. (2008). *Performance Practice and Process: Contemporary (Women) Practitioners*. Basingstoke: Palgrave Macmillan.

Brewer, M. (2005). *Staging Whiteness*. Middletown: Wesleyan.

Chomsky, N. (1993). *Year 501: The Conquest Continues*. Boston, MA: South End Press.

Deleuze, G. and F. Guattari. (1987). *Thousand Plateaus: Capitalism and Schizophrenia*. Minneapolis: University of Minnesota Press.

Eakin, P. (1999). *How Our Lives Become Stories: Making Selves*. Ithaca, NY: Cornell University Press.

Ellis, C. (2004). *The Ethnographic I: A Methodological Novel About Autoethnography*. New York: Altamira Press.
Foucault, M. (1980). *Power/Knowledge: Selected Interviews and Other Writings 1972–1977*. London: Harvester.
Freeman, J. (2007). 'Making the Obscene Seen: Performance, Research and the Autoethnographical Drift', *Journal of Dramatic Theory & Criticism* (Spring): 5–14.
Fusco, C. (1995). *English is Broken Here: Notes on Cultural Fusion in the Americas*. New York: The New Press.
——. (1999). *Corpus Delecti: Performance Art of the Americas*. London and New York: Routledge.
——. (2001). *The Bodies That Were Not Ours: And Other Writings*. London and New York: Routledge.
Gómez-Peña, G. (1994). *Warrior for Gringostroika*. Saint Paul: Graywolf Press.
——. (2000). *Dangerous Border Crossers: The Artist Talks Back*. London and New York: Routledge).
——. (2004). 'In Defence of Performance Art', in *Live: Art and Performance*. Ed. A. Heathfield. London: Tate Publishing.
——. (2005). *Ethno-Techno: Writings on Performance, Activism and Pedagogy*, London and New York: Routledge.
Gornick, V. (2001). *The Situation and the Story: The Art of Personal Narrative*. New York: Farrar, Straus & Giroux.
Gilmore, L. (2001). *The Limits of Autobiography: Trauma and Testimony*. (Ithaca, NY: Cornell University Press.
Heddon, D. (2008). *Autobiography in Performance: Performing Selves*. Basingstoke, Hampshire: Palgrave Macmillan.
Kaplan, C. (1992). 'Resisting Autobiography: Out-Law Genres and Transnational Feminist Subjects', in *De/Colonizing the Subject: The Politics of Gender in Women's Autobiography*. Ed. S. Smith and J. Watson. Minneapolis: University of Minnesota Press.
Kirby, M. (1969). *The Art of Time: Essays on the Avant-Garde*. New York: E. P. Dutton.
Kuhn, Thomas. (1962). *The Structure of Scientific Revolutions*. Chicago: University of Chicago Press.
Leguizamo, J. (2007). *Pimps, Hos, Playa Hatas, and All the Rest of My Hollywood Friends: My Life*. New York: Harper.
——. (2008). *The Works of John Leguizamo: Freak, Spic-o-Rama, Mambo Mouth, and Sexaholix*. New York: Harper.
Lejeune, P. (1989). 'The Autobiographical Pact', in *On Autobiography*. Ed. P. Eakin. Minneapolis: University of Minnesota Press.
Martin, J. (1990). *Voice in Modern Theatre*. London and New York: Routledge.
Owens, A. (2007). *The Magic Bullet: Metaphors for thinking through Applied Drama*. Practice Keynote Address, Drama Education 16th World Congress, IATA, Scheinling Castle, Austria (unpublished).
Read, A. (1993). *Theatre and Everyday Life*. London and New York: Routledge.
Reed-Danahay, D. (2007). 'Autobiography, Intimacy and Ethnography', in *Handbook of Ethnography*. Ed. P. Atkinson et al. London: Sage.
Russell, B. (1991). *History of Western Philosophy*. London and New York: Routledge.
Schechner, R. (1988). *Performance Theory*. London and New York: Routledge.
Slater, L. (2000). *Lying*. New York: Penguin Books.

Smith, S. and J. Watson (2001). *Reading Autobiography: A Guide for Interpreting Life Narratives*. Minneapolis: University of Minnesota Press.
Sokal, A. (2008). *Beyond The Hoax: Science, Philosophy and Culture*. Oxford: Oxford University Press.
Stanton, Donna (1987).*The Female Autograph*. Chicago: University of Chicago Press.

10
La Fura dels Baus's *XXX*: Deviant Textualities and The Formless

Roberta Mock

In April 2003, the Catalan company La Fura dels Baus brought their production of *XXX* to London and unleashed a furore in the headlines of Britain's tabloid press: 'Art? This is nothing more than porn' (*Daily Mail*); 'Sex on Stage Storm: Raunchiest show ever hits UK' (*Daily Star*); '*XXX*: Is this play art or porn?' (*Daily Mirror*); '*XXX*-Rated: Anger at hardcore sex act in shock new play' (*The Sun*). Clearly, relying on the press releases of the company's local master publicist Mark Borkowski, combined with interviews with those members of the opening night audience who somehow missed the warning that 'this production contains extreme and explicit sexual themes and images throughout', these articles could barely conceal their disappointment that *XXX* was operating well within the law.[1] Furthermore, while its live sex may have *appeared* real, appearances can be deceptive. Almost every tabloid included the same quotation from Carlos Padrissa, one of the production's co-directors:

> This is not pornography. It is art. The sex in *XXX* is not real but virtual. It is all theatre. It looks like real sex and the actors are often naked but it is just touching and kissing, there is no penetration .... But our aim is to make it look as if it were real.
>
> (qtd in Bynorth, 2003: 25)

XXX is based on the Marquis de Sade's 1795 novel *Philosophy in the Bedroom* and La Fura employed a range of technological devices in order to construct a performance text written through and as somatic possibility. The production opened with several startling visual images, one of which included the projection of a written sentence, 'A better world is possible,' on an enormous screen. This text was apparently generated by a live performer who was situated centre stage between screen and audience, exercising her pelvic floor muscles by manipulating a vaginally-clenched light pen on a digital writing tablet. This *coup de théâtre* was consistent with the company's established approach to contemporary technologies; indeed, while waiting for the show to start, audience members were encouraged to text chat room-style messages that would be projected on the same screen. Just as they had employed texting in earlier productions, La Fura had been playing with robotics and ballistics as a creative language since *Tier Mon* in 1988 and in *XXX* audiences were treated to flying fuck-machines, rotating platforms and an ingenious portable dressing room that cracked open to reveal a brightly lit bidet. Such enthusiastic experimentation with props contributes to the identification of features that have come to define La Fura's work: that is, the use of a 'multidisciplinary scenic language based on [the] predominance of the body and technology; invasion of the intimate space of the spectator [so that] the audience becomes a fundamental part of the show, similar to the Greek choir; and a theme that focuses on the conflicts between the individual and the group' (La Fura dels Baus, 2007).

As Sharon Feldman's analysis of their 1994 production of *M.T.M.* illustrates, 'the relationship between what is live and what is reproduced is completely jeopardized, and La Fura dels Baus's "great theatre of the world" is revealed to be ... a place that can only create a hallucination of the truth' (1998: 471). Central to the production of both message and affect in *M.T.M.* was the company's use of video. The audience were subjected to:

> A visually exhausting whirlwind of video images. In general these images fall into two major categories: ready-made images that are conceived *a priori* and images that are conceived during the actual performance. The problem that is thus posed for the spectator is that of deciphering which images fall into which category. All sense of space, time, and objective reality is altered through the use of the video camera and video projections.
>
> (Feldman, 1998: 468)

The target of La Fura's critique in *M.T.M.* was the aesthetic packaging of experience by the mass media and they used digital technologies at least in part to parody assumptions of authenticity. Audience members were 'simultaneously offered both presentation and representation, the actual and the virtual, pure presence and mediatized presence' (ibid.: 468). However, there are many potential reasons why truth may need to be presented as hallucination and, in *XXX*, La Fura constructed a similar spectatorial experience for a substantially different purpose.

According to *The Observer*, 'When a guy with a handheld camera moves around the audience, he first projects their faces, but mingles these with shots of couples groping, undressing, pictures which you realise only slowly are from elsewhere' (Clapp, 2003). *The Sunday Telegraph* was disorientated, but admitted that *XXX*:

> Kept us concentrating – even when two men, one very muscular, added their naked limbs to the tangle on the revolving circular bed. Who was doing what to whom – and with what? The screen was no help: it fragmented into dozens of heaving mini-screens until you really couldn't tell what was a bottom and what was merely an armpit.
>
> (Herbert, 2003)

For *The Independent*, both live and digital bodies could be considered equally authentic and thus economically valuable: 'What was genuine (the actresses' cellulite-clad bottoms also had the ring of authenticity) were the video clips, which included a woman with a penchant for ponies; orgies; ... a nipple being surgically cut into; and a prolonged cum-shot. Not bad for 25 quid' (Stuart, 2003).

Unlike *M.T.M.*, *XXX* was no parody of mediatization. Rather, the inability to distinguish between the live and the digital – that is, the hybrid, mediatized experience itself – was required in order to 'authentically' embody the impossibility of fantasy. The distinction can be best summarized by comparing *M.T.M.*'s epilogue to the prologue of *XXX*. In the former, a sentence in Catalan was flashed onto a video screen; it read 'No em llanceu més merda a sobre' ('Don't throw any more shit on me'). *XXX* opened with a projection of an ass spurting a blob of runny shit at the camera/audience. If, as Feldman suggests, *M.T.M.*'s epilogue is an evocation of exhaustion and a cry for authenticity in a space now steeped in 'layers of excremental versions of the truth' (1998: 471), then *XXX* opens with a vision of excrement in a liminal space characterized by its authenticity and vitality.

In this chapter, I will consider La Fura's challenge to our understanding of authenticity through Georges Bataille's concept of the *informe* (or 'formless') in order to discuss deviance, both corporeal and theatrical, that evokes horror and desire through simulated immediacy. Bataille's 'formless' is a paradox in that it can only exist in relation to aesthetic form by signalling what it excludes. What Bataille, Sade and

La Fura share is the production of a meta-pornographic discourse in which obscene fantasy is recognized as the driving force of sexuality. Implicit in all three cases is that, while the realization of such fantasies – especially those that include rape, torture, mutilation, etc. – are intolerable, their imagination and representation should not be subject to state repression or censorship as this constitutes a denial of what makes us human. *XXX* performs the formless excess of Sade's novel in which, as in Bataille's theorizations, 'the individual loses their isolated subjectivity and joins with the loss "presented" by a work of art' (Hegarty, 2000: 144). La Fura bombards its audiences with a conflated pornographic-technological surplus that is inevitably rejected, thereby highlighting the difference between fantasy and reality as well as pornography and sexuality. As the *Time Out* critic noted of *XXX*, 'So many grinding genitals (magnified on an upstage screen) begin to seem much of a grimly mechanical muchness' (Logan, 2003).

Many commentators consider Sade's *Philosophy in the Bedroom* to be his most lucid and coherent political statement. Despite its recontextualization from late eighteenth to early twenty-firstccentury, and the omission of small (but significant) details, La Fura re-presents Sade's vision with remarkable accuracy and faithfulness. Very briefly stated, the narrative of *XXX* concerns the sexual (and therefore political and philosophical) initiation of an 18-year-old virgin, Eugénie (who was 15 in Sade's book) by a team of incestuous brother and sister porn industry veterans and their accomplice, Dolmancé. This education culminates in the willing girl raping her repressive mother with a strap-on dildo and then crudely stitching her vagina together. Sade's original version augmented this final scene rather more horrifically: Following penetration by her daughter, Eugénie's mother is raped vaginally and anally by a syphilitic valet and then sewn up so none of the poison can escape. We also discover that it is Eugénie's father who has sanctioned and arranged the education of his daughter and licensed the treatment of his wife who would inevitably come to fetch her.

Sade's libertines (both male *and* female) conflate sexual and political power. As Angela Carter proposes in her classic study, *The Sadeian Woman*, from the inverted axiom 'I fuck therefore I am':

> [Sade] constructs a diabolical lyricism of fuckery, since the acting-out of a total sexuality in a repressive society turns all eroticism into violence, makes of sexuality itself a permanent negation. Fucking, says Sade, is the basis of all human relationships but the activity

> parodies all human relations because of the nature of the society that creates and maintains those relationships.
>
> (1979: 26)

Sade's heroes and heroines, according to Carter, 'fuck the world and fucking, for them, is the enforcement of annihilation'. La Fura's publicity states that they attempted to make 'no moral judgement' on the story's narrative. Alex Ollé, *XXX*'s co-director, recognizes that most commentators on Sade's work find at least one system of morality inherently at work in his texts, stating in an interview before the production came to Britain that 'Sade takes you to a point where you are forced to reject him. Actually, I think he is a bit moralistic' (qtd in Tremlett, 2002). Presumably, the omissions I outlined in the final scene (that is, syphilitic inoculation, the treatment of wives as chattel, the implications of incestuous paedophilic abuse, etc.) represent the point at which La Fura rejected Sade. It would perhaps have been clearer to say that La Fura were making no moral judgement on the Sadean universe they felt able to recreate under the circumstances, refusing to indicate to audience members precisely how they should react to what they were experiencing. The questioning of morality was left to the individual; its tolerable limits (that is, the definitions of taboos and the identification of where and how transgression takes place) were not positioned by the company on our behalf. Perhaps even more accurately, La Fura publicly situated their audiences in the same space, with the same dilemma, as Sade's private readers.

It was Georges Bataille's generation that resurrected and rehabilitated Sade, that held a public séance and first made him perform on their behalf, both as justification for, and as illuminating contrast to, their own concerns and beliefs. But more so than any of his contemporaries, Bataille has been identified with Sade – occasionally simplistically identified *as* a reincarnated Sade – due to his methodologies, subject matter, obsessions (with waste products, for example), his materialist atheism, and the correlative relationship he proposed between eroticism and death. Each produced both 'clandestine' and 'signed' writings. Each were accused of either advocating fascism or providing highly sophisticated critiques of its dangers and repugnance. This resonates with contemporary criticism of La Fura dels Baus; in 1996, for instance, the Spanish theatre critic Enrique Centeno referred to the company as neo-Nazis and described their excessiveness as serving nothing more than 'an aesthetic and idolatry of violence and irrationality' (qtd in Breden, 2005: 276).

Michael Richardson points out that like Sade, Bataille 'believed writing should be thrown down as a challenge to the reader; it should be a deliberate provocation, and not serve a one-to-one relation in which the reader assimilates a message from the author' (1994: 16). I think it is reasonable, here, to substitute 'spectator' for 'reader' and 'performance-maker' for 'author' in order to consider performance works like *XXX* that are inspired by their writings. Famously, Bataille wrote that 'eroticism is the assenting to life even in death' which, as Richardson notes, would be unthinkable to Sade, for whom 'sex served to annul death'. For Bataille, it is during sex that we strive for our limits, the impossible. Our identities merge with our partners; we are lacerated and we collapse into a state of undifferentiated otherness. The Sadean sexual encounter avoids such contaminated otherness as it weakens our assertion of sovereignty as isolated beings. Bataille's notion of sovereignty and freedom, on the other hand, necessarily includes engagement and communication as a social being. For Bataille, a debauched sexuality – that is, one simultaneously excremental and sacred – was at the heart of mystical experience.

Two of Bataille's earliest essays, 'The Use Value of D. A. F. de Sade' and 'Formless', were probably written in 1929 and their connection is relevant here, in the nexus that is *XXX*. Yve-Alain Bois sees Bataille's notion of the *informe* (or formless) as 'nothing in and of itself'; it 'has only an operational existence: it is a performative, like obscene words, the violence of which derives less from semantics than from the very *act* of their delivery' (Bois, 1997: 18). The naming (or rather, un-naming) of this operation is of significant value in understanding La Fura's hybrid constructions of imaginative space and sensual textualities.

La Fura chose to devise *XXX* from a text that was performative but not a blueprint for performance (that is, not a script or playtext). As in Bois's explanation of the formless, I am referring to performativity in the Austinian sense of 'doing things with words'. Sade's *Philosophy in the Bedroom* was written in the form of a philosophical dialogue, not to be enacted but read. Sade, of course, considered himself a dramatist and at least one of his plays was professionally produced during his lifetime. However, as Austryn Wainhouse and Richard Seaver note:

> It is an understatement to maintain that, were his seventeen plays all that history had bequeathed us of his writings, Sade would hardly have a claim to immortality. The force, and indeed the essential worth, of Sade's works varies directly in proportion to their clandestine nature. The more open and public they are, the more conventional they

> become. The dramatic works, being most public, suffer most from conventionality and from what appears to be Sade's inherent timidity when faced with the dramatic form.
>
> (1989: 682)

Unlike Sade, Bataille did not leave us with any performance scripts, although apparently he wrote at least one which was later lost: it was a film script, never made, about a soap manufacturer who likes to pretend to be the Marquis de Sade and engages in the practices described in the *120 Days of Sodom* and *Philosophy in the Bedroom*. Not unlike La Fura, Bataille believed that this film would be 'commercial' (Surya, 2002: 349). *XXX* attempts what Sade and Bataille were unable to achieve themselves: that is, to transpose and translate the essence of both form and content to a visual embodied aesthetic medium. In the process it destroys the separation of these two terms, inscribing the operation of the formless.

I am borrowing my understanding of the term from visual art commentators Yve-Alain Bois and Rosalind E. Krauss, who curated an exhibition in 1996 entitled 'L'Informe: Mode d'emploi' and organized it through the four operational forms of the formless that they had identified: horizontality, base materialism, pulse and entropy. Broadly speaking, their usage derives from a kind of summation and distillation of Bataille's pre-Second World War writings, including his essay on Sade. Bois and Krauss both describe the formless as a process of declassification: 'in the separations between space and time (pulse); in the systems of spatial mapping (horizontalization, the production of lower-than-low); in the qualifications of matter (base materialism); in the structural order of systems (entropy)' (Krauss, 1997: 252). This declassing is closely related to processes of 'deviance' – that is, the generation of waste and monstrosity through the making of the ideal. All of the separations listed by Bois and Krauss are at work in *XXX*, a production for which the anus is a central motif and that features ecstatic wallowing in food and drink (spaghetti, tomato sauce, champagne and, most resonant of Bataille's own fixations, eggs).

I am using the formless, like Bois and Krauss, in an attempt to describe a process or an alteration that replaces semantic registers with an interpretive grid (Bois, 1997: 18). Admittedly, Bataille's own short essay on the formless is less than helpful in establishing a methodology:

> A dictionary begins when it no longer gives the meaning of words, but their tasks. Thus *formless* is not only an adjective having a given meaning, but a term that serves to bring things down in the world,

> generally requiring that each thing have its form. What it designates has no rights in any sense and gets itself squashed everywhere, like a spider or an earthworm.
>
> (Bataille, 1985: 31)

Like most of his writing, it is witty, contradictory, provocative. Bataille's writing on the formless ultimately implodes on itself; in this sense – like La Fura's *XXX* – the essay is what it describes. Bataille insisted that aesthetic value lay in immediacy, that to reach the purest form of ecstasy we must elude concern for the next moment and equally all those that follow after (Bataille, 1994: 91). *XXX* entropically removes objects and actions away from productive activity in the 'interest of the instant itself' (ibid.: 148), simultaneously destabilizing both space and time. La Fura created the illusion that the non-penetrative sexual actions we were watching live on stage were actually the scenes of penetration we were watching on the screen behind them. However, this penetration *did* take place, although in a different time, in the service of the moment.

The production's visual vocabulary and scenographic imagery – such as Eugénie's whipping by Dolmancé from his flying metal sex machine, her immersion in a translucent pod of water as she communicates through an internet chat room, and group sex on a trapeze – are probably best appreciated by audiences that are used to reading bodily extremity, immediacy and technological innovation as creative languages in themselves (as opposed to those acting in the service of another language). To be more specific, performances like *XXX* need to be read experientially in terms of the slippage between our understanding of actuality and illusion, rather than through the recognition of theatrical action as metaphor. This means that audience members must allow themselves to fall into the gaps where actuality and illusion can never quite meet – the spaces Bataille might call 'the impossible' where we are both free and powerless.

Using his theorization of two impossible forms of communication, Bataille enables us to read certain types of performance event in terms of sacrificial poetics. Although he (thankfully) notes their very different levels of modern social acceptability, he sees in both human sacrifice and poetry a close correlation: 'both a sacrifice and a poem withhold life from the sphere of activity; both *bestow sight* on what, within the object, has the power to excite desire or horror. ... poetry is no less directed toward the same aim as sacrifice: it seeks as far as possible to render palpable, and as intensely as possible, the content of the present moment' (Bataille, 1994: 149, emphasis in original). In *XXX*, there are distinct references to sacrifice, as well as echoes of Bataille's parodying of Christianity which he sees as a force for denying human spirituality in its refusal to admit ecstatic ritual as well as its belief in a sovereign Godhead; an obvious example is a symbolic crucifixion generated through video projection. More importantly, however, it is a production that renders intensely palpable the hallucinatory quality of the present moment, consistent with La Fura dels Baus's performance strategy since the company's formation in 1979.

It was their 1983 production *Accions*, best described as site-specific performative installation, that made the company's agenda concrete; they described it as 'a game without rules, a drink thrown in your face ... a brutal stream of hammer blows, a sound execution, a chain of unlimited situations' (qtd in Saumell, 1998: 19). Since then, they have collectively created large-scale street and stadium spectacles (including part of the jaw-dropping Opening Ceremony for the 1992 Barcelona Olympics), internet-based projects, corporate events, and opera. In 25 years, *XXX* was one of relatively few productions staged by La Fura in traditional theatre buildings and only their third based on a literary text (the company refers to these as 'matrices literarias' or literary matrices). It is entirely inappropriate to locate La Fura's artistry in 'traditional' theatrical criteria – that is, in the depth (as opposed to the transparency) of the illusion; in the clarity and explicitness of interpretative opportunity; in the so-called Cartesian duality of mind and body. By these standards, La Fura's theatricality can only inevitably be considered deviant.

I am proposing that we should embrace such deviance as a way of aesthetically theorizing what Susan Broadhurst has identified as 'liminal' performance, those experimental forms and practices that are located on the thresholds of the physical and the virtual and at the interfaces of body and technology (2007: 1). This is perhaps especially valuable for performances that also explore the dynamics of sexuality. David Ian Rabey has identified the following characteristics of theatrical eroticism:

> the extrapolations of the imagination, the reflexive self-consciousness, the interrogation and suspension of notional 'reality,' the dialectic of presence and absence (and proximity and distance), and the stretching and challenging of time.
>
> (2008: 63)

These impulses correspond uncannily with those that operate in much digital performance and, more specifically, in La Fura's work.

It is probably no coincidence that the company included an early version of their 'Binary Manifesto' on the *XXX* website (which also included a number of interactive games):

> Within Digital Theatre the absolute abstract idea lives together with the return to the body, which may adquire [*sic*] a sado-masochist, sensual, angelic, or orgiastic dimension, or perhaps even a mixture of all of them. A cyber-body that also implies cyber-sexuality ... By definition, the theatrical act always offers an excess, an extra performance. It is the pleasure of showing, and showing off. A means of identification is established between the actor and the spectator ... the digital image speaks in the present: 'Ladies and Gentlemen, this is how it is.' This is why it can be twinned with the live act, with the theatre, and with the here and now. Therefore, XXX is an interactive proposal that allows the image to be mutated by making it pass from one format to another – virtual and real, by placing it on different stages.
>
> (La Fura dels Baus, 2003b)

The message on a more recently created company website is further refined. Now, digital theatre 'refers to a binary language connecting the organic with the inorganic, the material with the virtual, the actor in the flesh with the avatar, the present audience with the internet users, the physical stage with the cyberspace' (La Fura dels Baus, 'Binary Manifesto', n.d.). Here the virtual body is explicitly linked to both genitalia and sex/gender systems: 'Will digital theatre perpetuate Phallocracy? Will Vaginocracy eventually win? Or will both join in perfect harmony 0-1?'

La Fura's Binary Manifesto owes an obvious debt to Donna Haraway's groundbreaking feminist Cyborg Manifesto: 'A cyborg is a cybernetic organism, a hybrid of machine and organism, a creature of social reality as well as a creature of fiction ... The cyborg is resolutely committed to partiality, irony, intimacy, and perversity. It is oppositional, utopian, and completely without innocence' (Haraway, 1991: 149, 151). Perhaps most relevant to La Fura's stagecraft in *XXX* is Haraway's observation that 'in imagination and in other practice, machines can be prosthetic devices, intimate components' (ibid.: 178). We are reminded of this when the Madame boasts that the sex toy modelled on her genitals has sold more copies than the bible (we are told that 'the deluxe version has an asshole'). As Mercè Saumell notes, in La Fura's combination of the atavistic and technological:

> The machine appears in the group as an object that interacts with the human, that answers, executes, and satisfies his/her desires. The

> machine is figuratively sodomized into complying with human orders .... [*XXX*] flaunts the criteria of simulation as characteristic of the erotic (against the real act flaunted against pornographic works).
> (2007: 344)

Not only are the bodies of the actors extended temporally and spatially via video projection, so too are they shaped materially through prosthetic devices. When a 'volunteer' from the audience allowed Eugénie to fellate him, astute audience members probably noticed that his flaccid penis seemed rather independent. Still, it was this 'plant' and his actions – was he? wasn't he? will he? did he? will you? will you have to? – that most excited the British newspapers.

Experience has shown us (for instance, in the case of the *Romans in Britain* trial) that the tabloid media often plays a crucial role in the demonization of certain performances. Ironically, however, in tossing the bait to the supposedly slathering masses (who chose not to take it), the tabloid journalists who had not seen *XXX* seemed to express its significance far more astutely than the broadsheet critics who had. According to Wendy Steiner:

> Pornography and pornographic art are important because they mark the bounds between thought and deed, and like every such liminal zone they are fraught with fear – fear that fantasies will come true, will invade the world of public action – and the opposite fear, that there will be no such crossover, that the pleasure and energy and justice of this zone will have no realization outside it.
> (Steiner, 1995: 38)

This is a debate with which the 'serious' critics failed, or perhaps refused, to engage and yet seemed to suggest if one paid close attention to the sublimated leakings of their collective text. *XXX* was described as both too simulated and too stimulating; it was boring and yet there was too much going on; it was too distanced and yet the audience was too forcefully involved; it was repetitive and yet contained a string of remarkable – or gratuitous – images and visual effects; it was both tired and exuberant, witty and witless, reactionary and too naïvely literal. Over and over again, the critics referred to what they personally *experienced* and *felt* but this, somehow, wasn't considered *enough*. I have never seen reviews in which the critics pushed themselves so far to the front of the writing – and their seemingly unintentional exposure, perhaps, is why they needed to protect themselves by ultimately rejecting *XXX*. In discussing this production, the critic was forced to open up, to publicly recognize her fantasies and limitations.

The critical responses to *XXX* mirrored those generated by Sade's original writing – which the vast majority of reviewers had clearly never read (but alluded to as if they had). The startling visual image which opened the production was considered spectacular but either empty or confusing. Admittedly, the Madame's vaginally scrawled reminder that 'A better world is possible' is an extremely truncated version of the fifth section of *Philosophy in the Bedroom* entitled 'Yet Another Effort, Frenchman, If you Would Become Republicans' (which was detached from the novel and published as a patriotic pamphlet during the Revolution of 1848). But those who know Sade's life and work also know that he, like Bataille later, loved revolution for the revolt itself. My point here is not that audiences *should* know this but that the specific political point expressed by La Fura, through a visual cyborgian textuality, was interpreted based on assumption rather than signposts erected by the company. Who ever said, as most critics presumed, that the production was supposed to exclusively, or even primarily, titillate and that it should be considered a failure if it didn't? *The Guardian*, taking a seemingly more sophisticated attack route, first objected to La Fura's 'unquestioning acceptance of Sade's dubious philosophy' and then blamed the company for not providing us with 'the genuine dialectic' of Peter Weiss's *Marat/Sade* (Billington, 2003). What I believe was *really* being critiqued was not the company's supposed message but the manner in which it was 'declassified' by La Fura's performance strategy.

As Yve-Alain Bois notes in his discussion of the formless, 'It is neither the "form" nor the "content" that interests Bataille, but the operation that displaces both of these terms' (1997: 15). Similarly, the doubling of corporeal deviance by means of theatrical deviance thereby depositioning audience members experientially and morally, rendered *XXX* an affective operation rather than a product. According to Bataille, Sade's imagination of excess and doctrine of irregularity retains value for an individual who 'knows that he must become aware of the things which repel him most violently', since 'those things which repel us most violently are part of our own nature' (Bataille, 1986: 196). I suspect that, like Bataille, La Fura were more interested in discussing Sade 'with people who are revolted by him' (Bataille, 1986: 180). *XXX* was an attempt at a conversation that required of its audience a complex but open engagement with theatrical presence. Its hybrid textuality fused technology, physicality, and many forms of writing (from Sade's 'philosophical dialogue' to vaginal mark-making to breathless preview articles in tabloid newspapers), in order to embody what lurks at the very edges of collective imagination.

Note

1 In Australia, in February 2004, the pre-recorded video elements of *XXX* were subject to censorship. Having agreed to pixelate the most graphic imagery, the production (or, more specifically, the filmed elements it included) received an R18+ (Strong Sexual Violence, Sexual Activity) rating, meaning that people under 18 years of age would not be admitted. See <http://www.refused-classification.com/Films_furadelsbaus.htm>, accessed March 2008.

References

Bataille, Georges. (1985). *Visions of Excess: Selected Writings, 1927–1939*. Ed. Allan Stoekl. Trans. Allan Stoekl with Carl R. Lovitt and Donald M. Leslie Jr. Minneapolis: University of Minnesota Press.

——. (1986). *Erotism: Death & Sensuality*. Trans. Mary Dalwood. San Francisco: City Lights Books.

——. (1994). *The Absence of Myth: Writings on Surrealism*. Trans. and intro. Michael Richardson. London and New York: Verso.

Billington, Michael. (2003) Review of *XXX*, in *The Guardian*, 25 April 2003. Reprinted in *Theatre Record*, XXIII.9: 545.

Bois, Yve-Alain. (1997). 'The Use Value of Formless', in *Formless: A User's Guide*. Ed. Yve-Alain Bois and Rosalind E. Krauss. New York: Zone Books.

Broadhurst, Susan. (2007). *Digital Practices: Aesthetic and Neuroesthetic Approaches to Performance and Technology*. Basingstoke: Palgrave Macmillan.

Breden, Simon. (2005). Review of *La Fura dels Baus: 1979–2004*, in *Contemporary Theatre Review*, 15.2: 275–7.

Bynorth, John. (2003). 'Sex on Stage Storm: Raunchiest show ever hits UK', *Daily Star*, 24 April, pp. 1, 25.

Carter, Angela. (1979). *The Sadeian Woman: An Exercise in Cultural History*. London: Virago.

Clapp, Susannah. (2003). Review of *XXX*, in *The Observer*, 27 April. Reprinted in *Theatre Record*, XXIII.9: 544.

Feldman, Sharon G. (1998). 'Scenes from the Contemporary Barcelona Stage: La Fura dels Baus's Aspiration to the Authentic' *Theatre Journal*, 50.4: 447–72.

Haraway, Donna. (1991). *Simians, Cyborgs and Women: The Reinvention of Nature*. New York: Routledge.

Hegarty, Paul. (2000). *Georges Bataille: Core Cultural Theorist*. London: Sage Publications.

Herbert, Susannah. (2003). Review of *XXX* , in the *Sunday Telegraph*, 27 April. Reprinted in *Theatre Record*, XXIII.9: 541.

Krauss, Rosalind E. (1997). 'The destiny of the Informe', in *Formless: A User's Guide*. Ed. Yve-Alain Bois and Rosalind E. Krauss. New York: Zone Books.

La Fura dels Baus. (2003a). *XXX*. Riverside Studios (London), 17 May. Dramatized by Mercedes Abad, Alex Ollé, Carlos Padrissa and Valentina Carrasco; directed by

Alex Ollé and Carlos Padrissa; performed by Teresa Vallejo (Madame), Pau Gómez (Giovanni), Sonia Segura (Eugénie) and Petro Gutiérrez (Dolmancé).

——. (2003b). *XXX* website. <http://www.furaxxx.com/xxx/english/menu/menu.htm>, accessed September 2007.

——. (2007) *Imperium* website. <http://www.imperiumlafura.com/en/synopsis/>, accessed July 2008.

——. (n.d.) 'Binary Manifesto'. <http://www.lafura.com/entrada/eng/manifest.htm>, accessed June 2008.

Logan, Brian. (2003). Review of *XXX*, in *Time Out*, 30 April. Reprinted in *Theatre Record*, XXIII. 9: 544.

Rabey, David Ian. (2008). 'The Theatrical in the Sexual, the Sexual in the Theatrical: Some Parallels and Provocations', *Essays in Theatre/Etudes Théâtrales*, 21.1–2: 63–77.

Richardson, Michael. (1994). 'Introduction', in Georges Bataille, *The Absence of Myth: Writings on Surrealism*. Trans. and intro. Michael Richardson. London and New York: Verso.

Saumell, Mercè. (1998). 'Performance Groups in Contemporary Spanish Theatre', trans. Jill Pythian and Maria M. Delgado. *Contemporary Theatre Review*, Volume 7, Part 4: *Spanish Theatre 1920–1995: Strategies in Protest and Imagination (3)*.

——. (2007). 'La Fura dels Baus: Scenes for the Twenty-First Century', trans. Simon Breden, Maria M. Delgado and Lourdes Orozco. *Contemporary Theatre Review*, 17.3: 335–45.

Steiner, Wendy. (1995). *The Scandal of Pleasure*. Chicago: University of Chicago Press.

Stuart, Julia. (2003). Review of *XXX*, in *The Independent*, 29 April. Reprinted in *Theatre Record*, XXIII.9: 544.

Surya, Michel. (2002). *Georges Bataille: An Intellectual Biography*. Trans. Krzysztof Fijalkowski and Michael Richardson. London and New York: Verso.

Tremlett, Giles. (2002). 'More Sex Please, We're Spanish', *The Guardian*, 5 June. <http://www.guardian.co.uk/culture/2002/jun/05/artsfeatures.culturaltrips>, accessed April 2008.

Wainhouse, Austryn and Richard Seaver. (1989 [1966]) Introduction to Part Four, in The Marquis de Sade, *One Hundred & Twenty Days of Sodom*. London: Arena Books.

11

Bodies in Suspension: The Aesthetics of Doubt in *Honour Bound*

Rachel Fensham

As with the bitterly ironic words, *Work Makes Man Free*, above the gates of Auschwitz, the gates of the prison at Guantanamo Bay are adorned with a sign saying, *Honour Bound to Defend Freedom*. From this text, theatre director Nigel Jamison took his title for a dance theatre work that explored the conditions under which Australian citizen David Hicks could be imprisoned at Guantanamo without trial. Hicks's case was one of the most prominent to expose the political and ethical gaps in human rights discourses under the new regimes of torture imposed by the US government on the post-9/11 global mediascape. Utilizing dance, digital technology and verbatim theatre, Jamison began *Honour Bound* with a simple proposition: he saw the image of 'this human figure spinning and turning in a void' and, with Garry Stewart, tried to activate the sensory deprivations of a body denied expression by what Giorgio Agamben has called 'a zone of exception' (Phillips, 2006).

Let me introduce the situation. Early in the 'War on Terror', Hicks was picked up on the borders of Pakistan and Afghanistan as a potential terror suspect. Having left Australia as a disaffected young man, and convert to Islam, he had spent time training (no one knows how long or how seriously) with the Taliban just as the United States announced the expansion of its war efforts in Iraq. His capture, along with that of other potential suspects in the wrong place at this time, led to his 'extradition' via Egypt to Guantanamo Bay. Serious allegations of torture conducted during these journeys between states continue to be debated in the world's press but the political no-man's land of the camp was a more permanent horror.[1] In this detention camp established on an American base in Cuba, the world has seen glimpses of the animal cages where suspects were detained, the orange overalls of the prisoners and their shackles. As a white man deemed sympathetic to the Taliban, Hicks became a key figure in the US campaign to determine and exemplify the 'invisible' threat that their nation faced. In 2004, he and his military counsel, Major Mori, produced an affidavit alleging torture at the hands of his captors, a detailed statement which produced no response from the then conservative Australian government. In spite of growing legal and public concerns about physical abuse, his psychological state and the injustice of his detention, Hicks was to remain in limbo for more than six years without trial.

In this chapter, rather than detail the case, I want to focus on how the images of a body in suspension, entangled by legal, political and ordinary understandings of international justice, shaped the choreography and reception of this performance. For various reasons, the peculiar phenomenology of dance, with its elusive, often subjective meanings, resists definitive interpretation: the speed of choreographic dynamics rarely allows an oppositional stance to emerge and any direct commentary from the theatricalized dancing body requires careful crafting. This does not deny the potential efficacy of dance as an embodied politics; however, on those occasions when a choreographer presents a political work, as with William Forsythe's *Three Atmospheric Pieces* (2006), it can be poorly received – critics dubbed that work the 'Baghdad Ballet' (Mahony, 2006). Without many artistic precedents, *Honour Bound* was conceived as a dance-work that might mediate a political intervention from inside the fast techno-aesthetics of contemporary dance.

Recent interest in the affective dimensions of 'watching dance' suggests that another approach might be possible with which to consider a work such as *Honour Bound*, given that the complex sensuality of choreography make many modalities of perception active. Dance scholar Susan Leigh Foster, writing on 'kinesthetic empathy', has usefully proposed several concepts with which to theorize the powerful emotional and corporeal experience of dance audiences. Historicizing the differences in relations between choreographic form and spectator, she contrasts the sensory interpretation of dancing bodies in Renaissance choreography, modern dance, and dance constructed with new technologies of movement and representation. The structural patterning of Renaissance court dances offered its audience a 'spontaneous and mechanical' approach to a scenic or pictorial logic, whereas the abstract dynamics of modern dance have valorized the 'intrinsic intertwining of muscular action and emotion' (Foster, 2007: 249). Twenty-first-century dance forms provide less formal or psychological satisfaction to the viewer, although she suggests, they give 'fugitive, flickering glimpses of one another's corporeal status as it transits, blurred into the prosthetic devices that intensify even as they obscure physicality' (ibid.: 254). Without losing the intimacy, or sensuous responsiveness, to embodied subjectivity, she argues that new modes of dance produce a 'corporeal situatedness' that give the viewer a transitory yet sensory imprint of how kinaesthetic realities are experienced by different bodies.

Through the bodies of a team of circus artists, dancers and acrobats, I want to suggest that an imprint of the effects of incarceration, with the attendant loss of sensation and intimacy, and the experience of physical cruelty, were communicated kinaesthetically by *Honour Bound*.[2] Without imitation or naïvety, the production attempted to make sense of that which denies the sensibility of the subject. In order to deal with an experience that torture victims describe as being 'compacted in pain', the spectator therefore needed to feel circumscribed by the corporeal situation of suspension (Daniel, 1994). Suspension has a double meaning in this context, one that could be both formal and expressive, and another that might be binding and real; in political and legal terms suspension implies the cessation of rights, functions or privileges; however, at a sensory level the quality of suspension involves the feeling of hanging from a fixed point, or the making of a minimal movement about one point of attachment. This chapter is thus about suspension and its relationship to torture as a limit condition of imprisonment. Following a discussion of texts, technology and torture, I want to suggest that Stewart's choreography adapts and extends body art performance that has used the dynamics of suspension (Figure 11.1) to test the limits of pain. Witness to the disconcerting effects of flickering sensations, I will conclude by reflecting on the political and aesthetic demands of doubt.

Texts, technology and torture

From a swarming crowd of orange-clad figures, blinded by intense overhead search lights and the noisy sound of helicopter blades whirring, one figure is plucked out on a rope, swings up and disappears from view. Across the back wall of the stage appears the projected text of the Geneva Convention on Human Rights, written in 1948 after the Second World War, in the hope that such atrocities would never happen again. A deep masculine voice intones:

> *This declaration reaffirms the belief in the dignity and worth of human rights and persons. Article Five. No-one should be subjected to torture,*

> *degrading or inhuman treatment. Article Twelve. All are allowed to have equal treatment before the law.*

The contrast between the responsibilities enshrined in international legal covenants and the manipulation and distortion of language that takes place when an imprisoning authority renames a prisoner of war an 'enemy combatant' is evidence of a post-structural logics at work in the world. Michel Foucault and Giorgio Agamben have theorized, for instance, the enmeshment of texts and technologies in frameworks of power, such as the militarism of the nation-state and the punishments of the prison. Other scholars, such as Judith Butler, have explained how power is dispersed through the psychic forces attached to discourse. In the postmodern dramaturgy of *Honour Bound,* the spectator becomes bound by a multilayered prison-house of texts. The projected image of a cage submerged the entire stage within a grid; which also dissolved as words – the Geneva Convention, the White House memos authorizing the use of certain kinds of interrogation techniques, the news reports – scrolled up and over the walls. Theatre reviewer Alison Croggan wrote:

> *The memos are projected on the back of the cage like a long road of text, along which a dancer is running. The text flips and the dancer falls into an abyss of darkness; and then he begins to run and climb again, and again falls, and again, and again. As an image of the impact of the State on real bodies, I have never seen anything so cogent and so powerful.*[3]

Against a type face collapsing and stacking up on itself, the running of the solitary figure was futile. As disordered fragments of text tumbled around him, the projections converged to make language, cage and torture function as violent mediations of desubjectification. I was reminded by this sequence that the muscle memory of soldiers has been trained to overcome an inherent resistance 'to killing without inhibition', through its practice with computerized simulations. The painful excesses of dance training also produce a muscle memory which propels the acrobat as dancer into the air, along a rope that lets them to fall again and again. In these situations, the technology of the replay or repeat event does not always lead to a liberatory body, a becoming-physical, because it can also detach the body from thought, from affect and from sensation. But muscle memory also records disturbances and lingers in the body as post-traumatic stress. In the text and camp of Guantanamo, the texts and global media, the texts and the public sphere, the determination of causality has been entangled by its presentational immediacy, but in the performance screen-play the acoustic repetition of speech becomes interwoven with the rolling, twisted bodies and produces a recognition and recall of muscular traces.

The Secretary of State, Donald Rumsfeld, itemizes torture treatments with an alphabet. As three more figures try to scale the wall, the written text blurs into screen static; now six are suspended, dissolving into the flickering until we realize that each bent or hunched body depicts the shape of a letter: W – false flattery; X – isolation; Z – deprivation of sleep.

Those who speak out against torture and those who have been tortured. But want to be heard. Can be located by dominant discourse in the 'domain of the inexpressible'. Going beyond Elaine Scarry's (1985) work on 'the body in pain' and its relation to torture, Darius Rejali describes 'the language-destroying power of torture', in which the psychically individuated experience of pain under conditions of torture have not been sufficiently heard in the public sphere. According to Nina Philadelphoff-Puren, 'inexpressibility in this context thus means not that the detainee "cannot utter" the testimony, but rather that having uttered it, it somehow fails profoundly to take effect, to be taken up by the political community to whom it is addressed' (2008: 204). After a plea bargain in which he agreed not to speak about any 'mistreatment' by the United States, Hicks was released into Australian government custody. However, under the conditions of this agreement all his previous testimony about the conditions suffered during detention was formally designated 'unspeakable'. Criminal sanctions were designed to render his previous statements non-juridical, as Philadelphoff-Puren argues, but they also positioned him as incapable of any speech opposed to the US discourse on terror. 'Forced to speak' in order to 'say that we did not hurt you', he cannot name the aggression taken against unspecified persons by the political system (ibid.: 215).

A verbatim interview with Terry Hicks is broadcast in grainy close-up. He explains carefully how his son has been treated, and occasionally pauses to reflect on his words: The situation he is in – it's not a good situation. This comment has all the nuance of Australian understatement. A figure is curled on the floor in the cage on stage, and we can hear thin, rasping breathing.

For Jamison, the relationship between father and son identified a powerful counter-narrative to the official texts and the affective dimensions of paternity gave structure to small claims for human decency. As with the media and public campaign to 'free David Hicks', Terry Hicks was to represent the 'ordinary' Australian. His calm but persistent tone gave voice to the reason needed to cut through the deceptive rhetoric that had allowed American military authorities to determine individual destinies at Guantanamo. The filmic presence of both Terry and Bev Hicks in the performance registered in palpable contrast to the hurling bodies and crashing images of projected text. When Terry spoke, one craned to listen because his deep, slow drawl emphasized only what mattered from the letters between him and David:

> *Sorry for the inconvenience. I am held in prison, waiting to receive a message, I need your support. I'm sorry, please keep in contact, I love you a real lot. God willing I will see you again.*

Like the strayed black sheep, David is the son who is never forgotten, and while not forgiven without justification, is always loved by his father. In the collaboration with Terry Hicks, Jamison found the personal voice of the story that needed to be represented, and renarrativized, for the theatre audience. This part of the performance, like much verbatim theatre, relied on its parallel retellings in the local and the particular. Its tone of persistent melancholia also gave utterance to David's unspoken words, allowing him to be situated as the defendant giving evidence of a lived situation before the spectator.

Bodies ... in the air ....

The dancing bodies are more complicated; as the surrogates for human flesh, they are rendered unlike us by the disguise of orange suits, as if a kind of flying circus. Instead of anonymity leading to dehumanization, Stewart hoped to resensualize the corporeal dislocation that might be felt in prison, solitude and inexpressible pain.

Figures chained at wrists to mess grid, one facing another, spinning back and out kicking the wall surface, sliding up, trying to reach out and through to touch the other. Bodies lying on the floor, picked up and locked in a solitary crate, bent over huddling against the wall.

Stewart is the Artistic Director of the Australian Dance Theatre and is one of a generation of Australian artists who have rejected the release

Figure 11.1 Suspended Body, *Honour Bound*, 2006. Photo with kind permission of Adam Craven

techniques of postmodern dance, yet continue to use its rigorous interrogative approach to movement. His choreography also plays with technological devices, and includes dancing with robots, using techno-music, or other forms of digital or mediated representation. Since its inception, however, Stewart's aesthetic has been driven by a desire to make bodies move at the extremes of their physical limits: 'my work was very muscular and forceful with a preoccupation with a driving velocity' (2007). Exploiting the technical virtuosity of ballet to get bodies into the air, he manipulates the physical possibilities of speed and irregularity from break-dancing, capoeira and martial arts. Energized by the hyperphysicality of the body, Stewart also experiments with the formal trajectories of both horizontal and vertical planes: it is as if his dancers fly through the air with all the magnificence of a man turning over a horse in a rodeo. His attitude of pushing dancers to their expressive limits is similar to Canadian choreographer Elizabeth Streb, who began hurling bodies at the floor in the 1980s, and it is worth recalling that the 'techno-bodies' of Streb's extreme action pieces were criticized by feminist scholars for a loss of interiority (Daly, 2002: 131).

Although committed to the political project of drawing attention to Hicks's incarceration, Stewart conceived artistically of 'this piece as an interesting extension of my vocabulary because there is not just dancing on the floor but also dancing on the walls and the ceiling and through mid-air' (Stewart, 2007). Much of the choreography takes place in micro-seconds of time, beyond the counting routines, because it exists only in the instants between a twist or fall in mid-air. In the collaborative live photo-performance called *Held* (2004), made with photographer Lois Greenfield, the beauty of Stewart's dance is revealed in the still moving image captured by the camera as 'the trajectory of each pose or movement ... arrested and held in the body before being re-directed, or the body returns to earth' (Marshall, 2008: 191). By disconnecting movement from expression, content from gesture, and by resequencing disjunctive phrases of dance vocabulary, this dance language is non-linear, and intersubjective, shaped by the thrill of intense sensation. Even though the individual dancing body seems disaffected, perhaps hardened, the concentrated efforts of corporeal activity, following prompts and impulses, are keenly felt muscular (social) memories.

Couples are pushing each other, folding around, touching, turning bodies against one another. Flipping one arm over another, these are bodies reacting to impulses even without pressure, or contact. They keep repeating the turns, twists, falls, beatings and rolls, thus exposing the external actions as effects. We hear the voice of Bev Hicks:

> *The Government demonized him and only now are the media treating him more humanely. There was a posed photo in Kosovo, of David holding a gun, one of his mates sold it to the press. People are gullible.*

Since Stewart's choreography interrogates dance limits, he was also willing to confront the physical limits of interrogation and torture in confined circumstances. As he says: the 'raw and chaotic physicality is an indication of how the world is being upended and how these are strange and difficult times. Through supporting the continuance of Guantanamo Bay, we are supporting the use of torture and illegal incarceration' (Griffin, 2006). For the artists, *Honour Bound* was never about whether Hicks is guilty or not, but about his right, and the more general human right, of a 'prisoner of war' to a trial. But what the choreography offered was something that approached the parallel 'quality of experience' relating not to pain, but to this odd suspension of physical borders, and private moments of sensation (Manning, 2006). The objective after-shock of the flying bodies, rather than their subjectivity, was responsible for producing an experience that could render a kinaesthetic empathy akin to the suspension of justice under a torture regime.

Analysis of the choreographic aesthetics is constraining in this performance, and the work could also be framed by the history of suspensions in live art practices. Extreme body art, or the 'explicit body' to use Rebecca Schneider's (1997) term, shows itself as the object of art itself, giving representation to a lived surface in the present, rather than in a reorganized or prefigured time. Notorious performance artists, such as thrill-seeker Californian Chris Burden or British artist Stuart Brisley, have their Australian counterparts in artists, such as Mike Parr and Stelarc, whose experiments with durational performance events have resisted the inscription and codifications of the body. 'The sheer physical endurance of these events', as Anne Marsh notes, often tested 'the limits of the body's capacity to survive and they also test the limits of the psyche: how much pain can the subject endure?' (1993: 108). The testing of pain thresholds was extended particularly to the 'aggressive tensions' produced in a sensory response to extreme actions.

A female dancer is performing disorientation, twisting on herself. Is this emotional hate? Her body is turned back on itself, arms twisted behind, falling on knees, in hump position, pushing against her head, she is hearing sounds that enter the brain. Shaking, trembling. One body picks the other up, only to leave the figure more in the dark, hands reaching towards the light. There is a sense of being nearly broken. We hear a soft singing lament of Islamic music. I am not sure whether this is a moment of tenderness but the figure turns away.

Performance artist Stelarc, now most famous for interactions with technologies that render the body 'obsolete', began his career with a series of suspension pieces. In *sitting, swaying event for rock suspension* (1980), hooks were pinned through the flesh around his limbs and back and attached to ropes that were in turn tied to rocks on the floor.[4] In an intricate arrangement of rocks, twine and fish-hooks, the Japanese aesthetics suggested a denaturalized form as well as the effects of shock, finessed by the puncturing of the skin. Held up for introspection, the elemental body led to contemplation of a fragile human flesh. Donald Brook writes that 'the pain experienced by Stelarc in body suspensions where the skin was pierced by hooks to enable his body to be elevated, are just some examples of the way in which sensation was foregrounded' (Marsh, 1993: 56). The challenge of artistic perception in these events was 'not a matter of sensation-giving but of information-getting, since art is ideologically continuous with life' (ibid.: 56). On the other hand, Anne Marsh argues that these performances served a double function, since they showed the 'artist-as-hero presenting a spectacle using his own body, sometimes presenting himself as a shaman who can heal himself or the sick society in which he lives or both; on the other hand, the body becomes the object of torture and is abused in an act of would-be liberation' (ibid.: 101). Subjectivity is foregrounded through an objectivity whose transgression of the fleshy body transforms both perception and situation.

Three solos. Picking up, lifting like body bags, nose-diving into the floor surface, jumping as if weight suspension comes from any part of the body, pulled up by knee joints, or twisted back through the ankle, and from the buttocks. One part not following the other.

A more 'explicit' political body appeared in Mike Parr's (2002) performance called 'Close the Concentration Camps', a work critical of the desert 'camps' established by the Australian government to contain refugee movements. Parr has frequently abjected his own body through the performance of 'cathartic actions', such as severing a prosthetic limb, vomiting coloured dye, or setting light to himself with matches. In repetition of the extreme protest actions taken by asylum seekers unable to speak out about their incarceration, Parr stitched his lips together in a gallery context. Making visual the sutured lips, replicating the silent act of the hunger strike, Parr sat in a chair dribbling spittle with his hands tied behind his back. The corporeal referent was thus explicit; however, he also draped the walls with the instructions given to guards in the detention centres as well as his correspondence with political authorities. The tension in this performance between voyeurism and cruelty, complicity and silence, could not have been more pronounced. Because the performance usurped the power of the protester to make a scene, not least because it invited the spectator to witness the artist's spectacular pain, the event was heavily criticized. Performance critic Edward Scheer has argued in his study of Parr's recent politically inspired pieces that the 'purposeful suspension of the aesthetic frame around this work does not foreclose on the representationality of the event' (2004: 23).

Staring above. Shaking. Nothing to see. One man hugs or holds another, carrying him to one side, but he cannot stand up, he falls against the walls of the cage. The other is dragged off the limp body by another that locks him in a cage. The face of Terry Hicks speaks:

> *David is not allowed to see sunlight so they only take him out at night. He was kept in solitary for 18 months in cage that was in a shed, 6 foot by 9 foot, no exercise. When he got friendly with the guard, they removed him and put in overhead cameras, then he had no-one.*

The ethical dilemma provoked by the artist's action delineates a response to the effects of illegal detention because it also produces an affective community in the assembled spectators. In this sense, Scheer argues that Parr's detention centre/torture acts can achieve a 'sense of a shared vision, or a shared sense of outrage at what is enacted in their names by political representation' (Scheer, 2004: 24). The power of the aesthetic action with 'its repetition of shock and image' is not that of denotation, instead a synaesthetic response is activated 'where external perceptions come together with the internal systems of memory and anticipation' (ibid.: 26). In other words, an affective witness to the tortured body might be one that facilitates 'the perception of the precisely regulated suffering of the Other in our midst' (Scheer, 2008: 55).

Senses of spectatorship

The perspective a spectator brings to a performance by Stelarc, Parr or *Honour Bound* is, as Maaike Bleeker points out, not that of witnessing 'naked reality', rather comprehension arises from the privileged, and culturally determined, modes of representation which govern our senses (2008: 13). One could not watch *Honour Bound* without also acknowledging the textual and corporeal practices of mediascapes that have perpetuated and condoned new torture regimes and human rights violations in the recent excesses of the War on Terror. The same 'incessant movement of iterability' which curtails speech and restricts visibility makes it possible however to re-cite, and re-site, torture testimony. In this way, art practices can generate 'the possibility that testimonial materials will not be permanently vanquished' just as the bodies that survive inhospitable conditions do not entirely disappear (Philadelphoff-Puren, 2008: 216).

> *We were forced to recognize how their stories are suppressed and will continue to be stifled. However, this ultimately does not engage the audience ... as the performers remain just dancers... they do not become characters we can connect with except through the movements of their bodies.*
>
> (anon. 2007)

> *Even the converted sometimes have to be reminded that in accepting the unacceptable, we destroy the best part of ourselves.*
>
> (Gardner, 2007)

When the anonymous London critic exclaims, 'except through the movements of their bodies', his ambiguity seems pertinent to the intolerable, indeed unjustifiable, representation of torture. My own immediate response to the performance surfaced as a kind of numbing silence, not because of ignorance to the facts nor because of an immunity to artistic intervention in relation to refugees or human rights. But somehow this performance was asking its audience, as the 'converted', to acknowledge these events as corporeal rather than simply intelligible. When the choreography held the dancers above the floor, propelled them out and around the ropes and bars of a cage filled with words that limited their freedom of movement, I am arguing that the bodies were literally and metaphorically suspended. This suspension produced a bodiliness, in a sensuous reality, for the gaps between texts and torture. The representation of torture has been paradoxical because the effects of sensory negation, and its confounding relationship to pain seem incommensurable: since torture denies the expression of pain, art can never approximate this annihilation of subjectivity (Scarry, 1985). In *Honour Bound*, no attempt was made to represent torture, nonetheless the choreographed suspension helped to foster doubt through a muscular sensation that we experienced in the 'flickering situatedness' of the dancing.

Held in suspension were sensations, thoughts and confusing ideas that usually fall between the gaps as we struggle to make sense of power. How should we read official texts? How can we listen to ordinary parents? How might we follow the writhing, leaping, running, of falling bodies suspended in a cage? In sensing the suspension viscerally enacted by orange-suited figures, and caught between utterances, it seems that the audience had a peculiarly acute apprehension. Erin Manning writes about 'prehension as the way in which the "pastness of experience" – reality – creates the potential for future connections', and yet, there is an active futurity-in-the-past that 'assures an active linkage between perception and event that makes prehension intelligible' (Manning, 2006). From my perspective, *Honour Bound* did not render linkages intelligible, indeed the situation of Guantanamo Bay, with its illegal practices make this impossible and too reductive. But the performance could be *sensible* of the phenomenology of torture as that of a subjectless disembodiment.

An interrogation scene was set on the side of the back wall, inside the projected cage, by two dancers suspended horizontally out from the wall. In this scene, a chair on which one of the dancers 'perches' is not a 'sitting perception' given by the body's sensuous relation to the haeccity of the object. Rather the upside down chair is the point of instability upon which the torturer conducts the interrogation, and from which he attempts to extract information.

Information-getting, as art critic Brook also attests, names the lie that torture turns against the subject. In this dance work, the body turning on the chair was temporarily dislodged, and the substitute became an object of physical punishment ready to be kicked or punched. In a series of stop-gap perceptions highlighted by strobe effects, the moving bodies of guard and prisoner were visible but they most effectively allow a 'particular set of prehensions [to] converge into subjective form' (Manning, 2006). We remember that subjectivity can be destroyed by painful interrogation even when the media prevents us from seeing or hearing it. At the same time, the relentless and amorphous repetition of the dancers' gestures refuse to cohere; it is as if the singular moments repeat the experience of torture as anti-form, as it is non-aesthetic, just at it is 'unspeakable'.

I have argued that *Honour Bound* succeeded as political intervention because it skilfully resisted making the suspended bodies feel 'normal', or 'natural'. In reality, Hicks was suspended in a state that is not natural, because his detention took place outside the regime of signs of a naturalized citizen. Since Guantanamo had no recognizable system of law, his suspension was not relative to the normal conduct rules of the Geneva Convention, and temporally it existed in an 'interregnum', between states. The choreography could not inscribe this political condition, nor record it in time, as if it was making a permissible commentary on technologies of representation. But it could materialize the effects of suspension, a hanging about, an interrogative crumbling of points of fixity, of certainty. Through the weight of athletic bodies, I am suggesting that a certain prehension could experience overdrive, free form and denial. In doing so, the performance established a mode of doubt over judgment, (whose? ours? the Australian government? the Australian public or citizenry? the US government?) and refrained from forming an opinion. As Elaine Scarry writes: 'To have pain is certainty, to hear about pain is doubt' (1985: 13).

The antithesis of *Honour Bound* is that a sense of honour, in the newly defined conditions of detainment, rendition and isolation, relieves the public of a sense of responsibility, outside of duty (moral, legal or military). Hicks's story, however, rests on the breaking of the rules by Major Mori and the lawyers and activists who pressed for his release. In the mainstream media, there is now a reluctance to suspend judgment, to show equivocation, about the evidence of torture, and thus to reside in the political sensorium of doubt. During this period, most media outlets were avoiding responsibility for putting pressure on the government in spite of the increasing number of senior law and political figures calling for reviews of the military tribunals and growing public support for Hicks. Indeed, the difficulty of speaking out and protesting the actions of government is one of the culturally determined systems that has shaped art practices post-9/11. The Melbourne producer of *Honour Bound,* Michael Kantor, for instance, felt obliged to declare that his company was 'apolitical' (Usher, 2006). And although he wanted to defer judgment, perhaps on artistic grounds, his views supported the silent majority. By way of contrast, the Australian reviews of *Honour Bound* were widely enthusiastic, because many critics were relieved to acknowledge their own dissent by supporting artists taking issue with the violent betrayal of citizens.[5]

Where political texts can deny bodily life, the 'technogenesis' of *Honour Bound* could transmit the effects of corporeal suffering, by showing pain's abstraction kinaesthetically. In the 'flickering' mode of expression Foster identifies with twenty-first-century dance, the sensory deprivation and manipulations that shape the experience of torture were palpable. Most importantly, they allowed the 'doubt' of ordinary people, such as Terry and Bev Hicks, to 'make of these [events] what we will' (Croggan, 2006).

The aesthetics of this dance-theatre, as a form of protest, thus assured an affective witness to a political discourse that had yet to be effectively renounced. This suspension of disbelief also opposed the desensitizing to human objectification produced by the convergence of torture with texts under the guise of twenty-first-century wars.

Notes

1 In February 2009, for instance, the UK government was accused of suppressing evidence on torture in relation to an extradition case relating to Binyam Mohamed. In the United States, debates on the evidence of torture allegations at Guantanamo Bay have surfaced after President Obama suspended the use of military tribunals to try potential terrorist suspects. Cessation of potential human rights abuses was notably his first political act on 20 January 2009.
2 Performers include David Garner, Alexandra Harrison, David Mueller, Marnie Palomares, Brendan Shelper and Paul White, and rigger Finton Mahony.
3 This potent convergence of digital text and falling bodies was noted by most reviewers, including the Democrat politician Andrew Bartlett in his blog, 3 August 2006.
4 Other suspension works include the *Event for stretched skin no. 4*, 'the body (was) suspended vertically for fifteen minutes, upside down, by the insertion of eighteen hooks into the skin' and another in which he was suspended for 32 minutes as he was lowered down an empty lift shaft' (Marsh, 1993: 102).
5 Although its opening season had limited success, the production has subsequently toured to the Vienna Festival (May), the Holland Festival (June) and The Barbican (London, November) in 2007 and New Zealand in 2008. It was recognized by the Australian Dance Awards 2008 as the Best New Choreography of 2006.

References

Anon. (2007). Review of *Honour Bound*, in *The Londonist*, 16 November.
Bleeker, Maaike. (2008). *Visuality in the Theatre: The Locus of Looking*. Basingstoke and New York: Palgrave Macmillan.
Croggan, Alison. (2006). Review of *Honour Bound*, in *Theatre Notes*, 16 September.
Daly, Ann. (2002). *Critical Gestures: Writings on Dance and Culture*. Middletown, CT: Wesleyan University Press.
Daniel, (1994).
Foster, Susan Leigh. (2007). 'Kinesthetic Empathies and the Politics of Compassion', in *Critical Theory and Performance*. Ed. Janelle Reinelt and Joseph Roach. Ann Arbor: University of Michigan Press, pp. 245–57.
Gardner, Lyn. (2007). Review of *Honour Bound*, in *The Guardian*, 16 November.
Griffin, Kelly. (2006). Review of *Honour Bound*, in *Beat #1030*, 13 September.
Jamison, Nigel. (2006). Writer-director. *Honour Bound*. Choreography by Garry Stewart, score by Paul Charlier. Premiere at the Playhouse, Sydney Opera House, 29 July 2006 with transfer to the Malthouse, Melbourne, 13 September 2006. A Sydney Opera House and Malthouse Theatre Commission.

O'Mahony, John. (2006). 'Baghdad Ballet', in *The Guardian*, 28 September.
Manning, Erin. (2006). 'Prosthetics Making Sense: Dancing the Technogenetic Body', *fibreculture*, Issue 9. Journal <fibreculture.org>.
Marsh, Anne. (1993). *Body and Self: Performance Art in Australia 1969–92*. South Melbourne: Oxford University Press.
Marshall, Jonathan. (2008). 'An Aubergian Reading of Garry Stewart's Held', *Still/Moving: Photography and Live Performance*, 8: 180–208.
Philadelphoff-Puren, Nina. (2008). 'Hostile Witness: Torture Testimony in the War on Terror', *Life Writing*. 5.2 (October): 203–20.
Phillips, Richard. (2006). '*Honour Bound* director Nigel Jamieson speaks with WSWS', <www.wsws.org>, accessed 23 August 2006.
Scarry, Elaine. (1985). *The Body in Pain: The Making and Unmaking of the World*. Oxford: Oxford University Press.
Scheer, Edward. (2004). 'A Vast Field of Lyrical Aggression: Politics and Ethical Spectatorship in Recent Durational Art by Mike Parr', *Broadsheet: A Journal of Contemporary Art*, Contemporary Art Society of Australia, 33.2 (June–August): 23–6.
——. (2008). 'Australia's Post-Olympic Apocalypse', *PAJ*, 88: 42–56.
Schneider, Rebecca. (1997). *The Explicit Body in Performance*. London and New York: Routledge.
Stewart, Garry. (2007). *Interview with Rachel Fensham*. Ballet Rambert, London, November.
Usher, (2006).

12
Translation: Words < – > Movement < – > Bits

Dawn Stoppiello

> Microsoft Word Dictionary Definition: *Writing – Words or other symbols, for example, hieroglyphics, written down as a means of communication.*

When I reflect on the words used to describe the term 'writing' in the Microsoft Word Dictionary, I see just how open the meaning of this word is. The definition speaks about what we do with writing but it doesn't say precisely what writing is – leaving that as broad as possible, symbols that are written down somewhere (page, space, computer chip) as vehicles for communication.

As a choreographer, I use visual symbols to communicate. Significantly, my creation of physical movement begins with language and words. Informed by decades of journal writing, inventing movement, and working with computer systems, my creative practice is a fluid interchange of languages, a series of translations between words, movements and bits. Troika Ranch, the company I co-founded with Composer/Media Artist Mark Coniglio, develops and presents works that use a hybrid of disciplines and techniques. Given this, I find it important to note that for many years our creative process always began with words. A single word (for example, 'body') would be the impetus for an entire evening-length work (*Vera's Body*, 1998). In one work (*The Chemical Wedding of Christian Rosenkruetz*, 2000), the alphabet served as the basis of an algorithmic system used to generate choreography from texts. I frequently examine movement sequences as a playwright would, for the content and meaning of dances as if the dances were spoken monologues. Often our completed performance works will also include literal translations – writing, semaphore and projected texts that appear on stage for the viewer to read. Part of my long-term creative practice is an investigation of a language-based improvisation, which provides an open system that I've successfully transferred into a movement-based improvisation. The conflict that arises between the translation of languages – the written and the physical – is, for me, a creative fission. It is the directness of words that attracts me to them and it is the enigmatic that draws me to movement. A colleague once relayed to me that in his etymological conception of performance, dance began with movement and theatre began with text. For me both are necessary and transposable.

To 'write', in twenty-first-century terms, seems to mean so much more than what Webster originally conceived. This century provides us with a multiplicity of means for communication, the Internet being the obvious means that I am going to refer to here. With the onslaught of digital reproduction tools and with so many keyboards and perceptions that can touch and alter the initial 'symbol' sent out to the world, it is as if our 'writing' can be more democratic, less precious, more fleeting, more passionate, less planned, more improvised, more interrupted, more like a conversation with a very large, and often unseen audience/reader. I recently saw a discussion on *Charlie Rose*[1] about the fact that newspapers in Los Angeles and New York City have been letting go of their film, music and dance critics and that to fill the gap of critical writing, blogs and websites have been popping up all over.

The writers on these blogs and websites are just average people interested in the mediums on which they are writing. They don't have the credentials of the prescribed 'experts' that were previously writing for the respected newspapers; however, they have enough passion and interest to want to start and maintain a blog or website, and many readers who respect their opinions. Because of this lack of critical writing in the newspapers and the shift to the Web as a place to find information, the role of the 'expert' is being called into question and the opinions of many are being considered as valid and useful. I was recently exposed to Twitter, a social network where the only question is 'what are you doing?' and the answer must be 140 characters or less. I was then 'followed' by other invited or interested 'Twits'[2]. I find this fascinating because the point of Twitter is to be in conversation with lots of others. Just that. Spontaneous, down and dirty, and instant – reach out and touch and be touched. There is something immediate and self-governing about this kind of communication. Nobody is excluded. All can speak about a topic. Anyone can be an expert. I will come back to more about the importance of this later.

As a maker of live performance, I tend to work in conversation with a limited few, at the opposite end of the continuum of communication that I described as happening in places like Twitter. I work slowly and thoughtfully in a closed environment and with chosen experts. This is a very different process to the way I share my work outside of the studio and stage space at this point in my career. For the public sharing of my work I take advantage of the (relative) democracy and access of the Internet. I post videos on YouTube and have a MySpace and FaceBook page for Troika Ranch so that my work can be viewed, discussed and shared with the multitudes of interested people representing many walks of life, with varying expertise and perspectives. In essence, there are two kinds of writing that I engage in around my work; the making and the sharing.

The making. *A description of two Troika Ranch works where significant procedural shifts occurred in my process, each of which demonstrates different aspects of my translation of words to movement to bits*

I think in words. I feel in images. I get up and make movement from those words and images. A computer then touches those movements. Computers are numbers. There are already a few layers of translations of symbols going on here, a few kinds of communicative writings in progress. Language is born of action, witness, recording. What inspired the first pictures and turned them into words? I can only imagine that it was early man's desire to be connected to others and unlock the contents of his own mind and body. Symbols were first made to encapsulate complete ideas. We use single symbolic gestures that mean complete ideas – peace, fuck you, ok, and thumbs up – for example. And there are all those little 'emoticons' and ascii symbol pictures that provide us with a minimal way to add an emotional framework to the words we write in chats and emails. Neither word or movement or image can speak a whole.

I have always seen dance and word-based language as two sides of a coin, the expressiveness of each being in their context and presentation and not in and of their discreet parts. Meaning does not come from phonemes but from collections of phonemes that have been trained to mean something. There is no universal repetition of movements between makers of dance that allow a general public to grasp the exactness of an implied meaning. Therefore each choreographer is creating his/her own set of phonemes that are organized into a kind of language. In dance there is no solid signifier or signified (as in language) yet I believe these are always implied, and that audiences try to find and understand them. Equally important to the equation is that dance-makers can't help but provide them. In my opinion, humans always try to make meaning out of their experience, even when the experience is not necessarily trying to convey a specific story or emotion. We can't help it. We seek to communicate. We are not abstract. We can be poetic and can find multiple and discordant meanings in single events but we seek meaning constantly. We are narrative. We are the meaning of narrative.

My work with Troika Ranch over the past 20 years has been a series of narrative expressions made through a translation of symbols via words, via the body (gesture), via the projected image, via computerized systems and via the stage. Since 1989 my collaborator Mark Coniglio and I have been in quest of a hybrid language comprised of these symbolic elements. I realize now that I believe humans are already fluent in hybrid language. We experience our world on many levels at once through the myriad symbols that surround us and we decode those symbols with all of the senses available to us. As an artist I simply try to mirror the fact that we humans do this already. I attempt to assist in the *understanding* that this multiplicity and hybridization already exist in our communication with each other.

'Rapping' the body

As a concrete example of how the translation of words to movements to bits occurs in my work, let me go back to 1990. While living in Los Angeles, Mark and I studied 'performance art' techniques with theatre director Scott Kelman.[3] One of the exercises we learned is called 'Rap' and is a series of instructions that allowed me to continuously speak improvisationally for an extended time. While it took me some weeks to become comfortable with the idea of speaking with my mouth rather then my body, once I did, I found an enormous potential for directly tapping into my subconscious mind and applying what I found there to my artistic practice. I used Rap in its original form, as spoken language, in several Troika Ranch works between 1990 and 2000. I used the technique in two ways: improvised on stage and as a means to develop monologues, dialogues, projected texts and overarching themes during the creative process. In 2001, I had the insight that the 'train' that I was on with words while rapping was the same train that I was on in my movement improvisations. I began to apply the specific instructions used in 'Rap', which allowed me to speak for lengths of time, to my movement improvisation practice in order to understand, codify and teach what I was already doing.

The first time I used this technique extensively as the basis for choreography in a full work was in the creation of our work *Future of Memory*

(2003). I call this practice 'Instant Choreography' and the rules are as follows:

> **Instant Choreography elements**WHOLE BODY: Utilize the whole body while containing it and staying open to potential.
> ENERGY: Bring energy to it.
> CYCLE IT: Arbitrarily grab a small bit and cycle it. Explore it. When informed change via Shift, Transform or Develop*.
> BECOME IT: Arbitrarily or when informed, first person it. What is it like? What are you? Who are you? What are you doing? Explore that.
> SPEED IT UP: Arbitrarily or when informed, speed up your tempo without losing energy and commitment.
> SLOW IT DOWN: Arbitrarily or when informed, slow down your tempo without losing energy and commitment.
> RHYTHM: Seek the rhythm of what you're doing and play that.
> BREAK THE LINEAR LINE: Arbitrarily change the type/quality of movement you are doing. Give up on it.
> FOLLOW THE LINEAR LINE: Stay with the type/quality of movement that you are doing. Don't give up on it.

*Shift: An instant pop into another form. Transform: Let the cycle evolve into something new. Develop: Go deeper into the 'intention' of the cycle.

> **The four considerations**
> Dance Ability (all you know about dancing)
> Changes (shift, transform, develop)
> Qualities and states (textures, tones, imagined realities)
> You now (your life, your memory, reality)

It has always been clear to me that when using a sensory system such as the MidiDancer[4] that improvisation is crucial for the dancer to take full advantage of the media at her disposal and 'play' that media freely in the moment. In *Future of Memory,* each dancer was equipped with a MidiDancer and now I had a method for improvisation with the device that unified the creative individuality of the members of the company. The piece centred on the social need to memorialize, and how individual and collective memories fade and change over time. This theme lent itself to the constantly shifting elements of the improvisational practice, as we were literally *in the moment* – defining, making and altering our recollection of a moment and the work at large, as we moved through it.

I still use this system regularly as a pure improvisational practice and as a means to generate choreography. I improvise, videotape myself, bring the material into iMovie and then select the most interesting moments. I place the selected fragments next to each other and edit them together to make a new phrase for the dancers to learn. I do this because I am interested in the conversation I am having with myself in the moment of improvising. I do not like to plan or organize too much. I like to stay connected to my subconscious mind and let it come through my body. This makes it difficult for me to be a writer. My nature is one of spontaneity. I live and move in the moment, and I want to capture that essence. This desire, I think, relates to what online social and chat networks and streaming video web sites are also trying to tap into – this moment now.

'Looping' a phrase

I love the edit because it is humanly impossible to reproduce. The attempt at achieving an edit in real time/space movement causes, firstly, a disruption of reality in the mind and body and, secondly, a shift which forces the dancer to make an individual choice of how to physically get from one place to the next. The dancers bring themselves into the phrase in this moment in an unprepared, unexpected way. In our work there is also a second translation that occurs through the computer when a dancer's actions are observed and turned into a series of numbers via a sensory suit like the MidiDancer or with a camera tracking system such as EyesWeb[5]. The relationship between the mover and the media on the stage become linked in time, but there is no automatic linkage to what is being signified by either the movement or the resulting image or sound.

One of my tasks as a choreographer is to highlight the links between the movements and the imagery and sound that is conjoined with the movement. I often describe this connection as the *metaphoric* linkage, it is the fact that the mover and the media ARE linked that is the real message. The nature of the linkage itself must sit well within the overarching context of the work I am making. What I mean is that the nature of the linkage plays as big a part in the meaning as the images and movements themselves. If a dancer is being tracked by a camera, then there is a kind of surveillance going on that cannot be ignored in terms of content. If they are wearing a sensory suit, then there is the addition of the 'exoskeleton' to be taken into consideration. Each kind of system also requires a certain kind of choreography to make it function. I choose the kind of sensing system based on the kind of metaphoric meaning I want to add into the piece, side by side with the kind of choreography I want to see and use.

A language translation occurs when the movements sensed by the electronic systems become data that can be processed as bytes/numbers. The language of computers is numbers and only that. The most exquisite gesture is reduced to a string of zero's and one's and it is up to the artist to re-expand those two numbers back into meaningful images, sounds and connections. Troika Ranch has spent 20 years grappling with this phenomenon.

This string of translations occurs in all of my works, but in 2007 another layer of translation was added when Troika Ranch began developing our current work, entitled *Loop Diver*. In our initial discussions of the work, we decided to abandon our frequently successful process of using a word as impetus. For *Loop Diver* we wanted to begin with a physical process and let the meaning be born from that. Out of our rehearsals we developed a new three-step system that combines a number of languages and translations.

1. The dancers and I improvise using the *Instant Choreography* elements.
2. We video tape the improvisations, bring them into iMovie and put segments together to make phrases in exactly the way I described above.
3. Translation through technology: Mark created a special looping tool (Figure 12.1) inside his Isadora® software.[6] The tool allows me to bring the movement material into the computer and impose complex loops on it.

A loop is a construct of audio and film technology and has been expanded upon and made more easily usable by *digital* technology. A loop differs

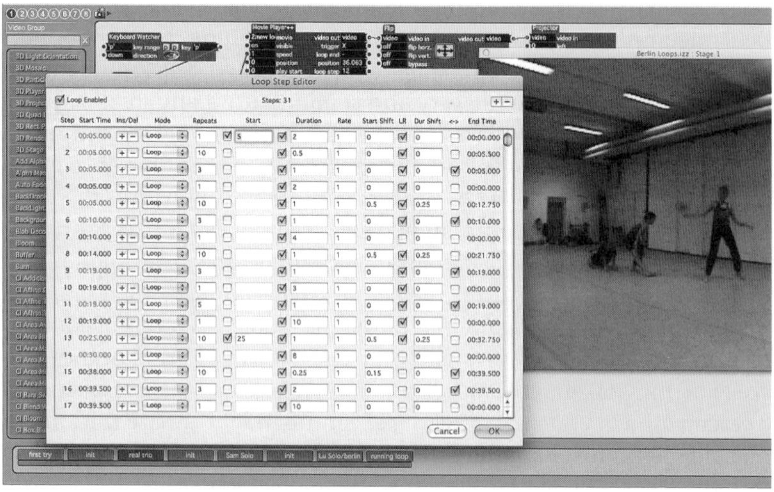

Figure 12.1 Screen shot of the Isadora 'loop editor'

from a natural cycle in that such cycles (for example, the seasons) always vary to some degree, and a loop does not. To illustrate a digital loop, imagine a recording of someone speaking the word 'machine'. A computer can generate loops from the individual sounds that make up the complete word in myriad ways; growing loops (m ma mac mach), shrinking loops (machine machin machi mach), sliding loops (mac ach chi hin), and so on. Once these loops have been applied to the movement material, the performers then attempt to learn the computer-generated looped phrases (Figure 12.2), a task that is daunting but critical to conveying the essence of the piece. These strict, unrelenting loops serve as a powerful metaphor for the loops of the mind, the prisons of repetition that occur when we experience a violent or traumatic event in our lives. Or even the simple result of living 40 years and having done a lot of things over and over again. By imposing the computerized looped material onto the dancers and pitting it against their inability to precisely perform the loops, the situation itself causes a stringent reverberation out of which they must dive and transcend. This is the theme that has become the cornerstone of this work. We could not have arrived at this thematic positioning without going through the rigors of the physical process first. Success.

A 'loop technique' is emerging from this process that demands a physical virtuosity based on several technical skills:

1. The Edit: a true edit requires no acceleration or deceleration on the part of the body – humanly impossible but the attempt provides an example of dealing with an external violent force.
2. Molecular Pause: a tiny but perceivable full body stop that exaggerates the end and beginning points of a looped moment.
3. Backwarding: identifying the reverse initiations of movement.
4. Speedshifting: identifying the various increments of potential speed and being able to move between them instantaneously.
5. The '4-I's' Improvisation: Imitate (try to loop like the computer), Internalize (put loop state in your mind/body/psyche), Inspiration (treat the loops you feel and see as a score for other movement invention), and Ignore (don't loop).

In addition to the technical aspects of looping, and because of the physical process, we have unearthed important psychological facets. After weeks of the dancers learning looped material from the computer, we realized that a psychological and physical 'violence' has been imposed. As in previous works, we again used Rap as a means of tapping into subconscious contexts and subtexts. In the dancer's solo monologue

Figure 12.2 Troika Ranch dancers Lucia Tong, Johanna Levy and Travis Steele Sisk learning looped choreography from the computer. Photo: Jennifer Sherburn

sessions each spoke about violence, trauma, anger, disintegrating bodies, not being able to breath, change/don't change, the flight of birds (a symbol of freedom), all leading to the deeper psychological content. After synthesizing these experiences into the overarching theme of violence, we discussed how this violence reverberates through the body and psyche until one has the realization, strength and fortitude to break the patterns that set in. Transcending the violence is the focus for developing the 'loop divers'. The loop divers are performers who, over time, and with increasing frequency, destroy the loops. They are the source of unpredictable turbulence that force change for the inhabitants of this world. Any performer can assume the role of a loop diver under a set of instructions with which to redirect the action. As in any kind of recovery process, a community can provide an environment to assist, but it is only the individual who can make the necessary break from pattern. The loop diver is the key to transcendence and change.

My next step choreographically is to explore the use of 'perpetual' motion in contrast to the intensely looped first half of the work. The interruptions of the 'loop divers' will intensify until the world shifts into a perpetually moving, non-repeating commotion of activity using some of my Instant Choreography improvisation elements. At first this shift feels like freedom and resolve, but over time it becomes clear that by never repeating an action we again find the performers in a kind of stasis.

Thus far we are not using any kind of sensory technology in *Loop Diver* (Figure 12.3) to track the dancers and manipulate media. Instead, the bits and bytes exist inside of the choreography itself – alongside the words and the movements. All encapsulated into one hybrid expression.

Now, more then likely, none of my elaborate processes are visible to an audience. They only see the physicality of what's in front of them. So, again, there a translation; my intention to their viewing. I appreciate and accept that individuals walk into the theatre with their own history and have a singular perspective from which to view my work. It is only in their mind that any real communication, and therefore meaning, happens.

Figure 12.3 Troika Ranch in *Loop Diver* (in-progress 2007). Photo: Oscar Sol

The sharing: *How I communicate my work to a larger, undefined community on the Web.*

Let me return to my conversation about Twitter and the Web. Of course, here I am writing for an academic anthology, a chosen set of experts. But usually, I share my work through vehicles like FaceBook and YouTube and, eventually, maybe even Twitter. As I have mentioned, the creation process of my work occurs in a closed space with my chosen experts and takes quite a long time. But how I choose to communicate my work to the public now includes the open, anti-expert, instantaneous platform of social networking sites. When Troika Ranch began to make *Loop Diver* we decided to draw on the Web for a kind of immediate experience. We wanted to reach more people, to get in-the-moment feedback from various kinds of people with various kinds of perspectives and expertise, and to join in the on-going dialog that is turned on even when your Internet Airport isn't.

We made a MySpace page for the project and began to blog about our process daily. We were in fact in the moment of *creating* the process. The blogging was a liberating experience because it allowed me to freely and immediately reflect on that day's progress. It was an improvisational approach to discussing my work. To fulfil my desire to stay in the moment as I do with my Instant Choreography, these writings were not thought over, edited or scrutinized before posting. Just posted and read and responded to by others. Several people asked deeper questions and requested explanations of some of the obscure and specific elements of our process. These questions caused us to come up with clearer definitions of the specifics and to reflect on the process of the moment. These interchanges were exciting and broadened our creative discourse. They brought more voices into our closed system, putting us into 'Instant Conversation' with a more general public and bringing the wisdom gained back into our process at the moment of being in it. The blogs also provided me with a documented reflection of my process that I could refer to days or months later. Not unlike the journals I had kept since I was a child, only now they were public and fluid. Looking back on those words nearly a year later brought me right back to that moment in the process which allowed me to approach the process of *now* with information from *then*, a translation from physical process to words (writing) back to physical process again.

In addition, we posted video clips on YouTube of *Loop Diver* as it progressed .We received many comments from the faceless audiences of the Internet that provided us with insights on how the material was perceived being tiny 'films' in their own right. Troika Ranch has not gone so far as to willingly put *all* our materials on the Web, but we are sharing clips of our performances and films and process more freely. In the beginning there was fear that our sacred, expert world was now open to scrutiny and plagiarism by the masses. When we initially put up a clip of a prototype of a film we were making, we felt a certain amount of ownership to the concept and techniques and didn't want any of it to be stolen from us. It would seem, at this late date though, that there are no new ideas, only one's individual take on or recontextualization of the ideas that already exist. At this stage of my career, it now feels more valuable to me to allow people who might never see my performances or films to be able to experience some version of them rather then keep them from experiencing any part of them. Who hasn't been copied? Didn't somebody once say that plagiarism is the highest form of flattery? And in the end, by putting Troika Ranch's work out on the web, we have received more feedback and praise and criticism than if we limited our audience to only those who had the privilege to get their butts into a theatre chair.

In summary, Troika Ranch performances, which themselves consist of a translation of words/movement/bits, now are 'written down' to the space of the Web to be, in turn, written about, blogged on, danced from, stolen, experienced and Googled again and again. As a society, we are all engaged in endless translation, a never-ending loop of transcribed symbols into multiple meanings. Ours is a hybridized language and one that we already speak fluently, despite transformations occurring instantly and constantly during this first part of the twenty-first century. Each mark we add to our alphabet is a symbol written down to communicate. These creations are what I love and work with – the twenty-first century's writings which are for me the language of the body, of technology and of art.

Notes

1 Charlie Rose: an acclaimed interviewer and broadcast journalist on American television.

2 'Twits' is my own term for the members of the online Twitter community. The Twitter community uses the term 'tweets' for the short messages sent between users.

3 Scott Kelman: American theatre director, <http://www.kelmangroup.com/scottBiog.php>, http://www.brooklynbay.org/, accessed 1 March 2009

4 MidiDancer; a sensory costume developed by Mark Coniglio in 1989. MidiDancer measures the flexion and extension of the major joints on the body and wirelessly transmits that data to a computer for use in live manipulation of media via the Midi protocol.

5 EyesWeb; a video tracking software developed by Antonio Camurri. <http://eyw.free.fr/feed/>, accessed 1 March 2009.
6 Isadora; software developed by Mark Coniglio, is a graphic programming environment that provides interactive control over digital media, with special emphasis on the real-time manipulation of digital video. <http://www.troikatronix.com>, accessed 1 March 2009.

References

Troika Ranch (Dawn Stoppiello and Mark Coniglio – Artistic Directors). (1998). *Vera's Body*. Set by David Judelson, Lighting by Susan Hamburger, Costumes by Katrin Schnabl. Performed by Mark Coniglio, Rose Marie Hegenbart, Lisa Herlinger Thompson, Dawn Stoppiello, Michou Szabo and Sandy Tillett. Joyce Soho, New York (April).

——. (2000). *The Chemical Wedding of Christian Rosenkruetz*. Lighting by Susan Hamburger, Costumes by Katrin Schnabl. Performed by Mark Coniglio, Danielle Goldman, Anthony Gongora, Lisa Herlinger Thompson, Dawn Stoppiello, Michou Szabo and Sandy Tillett. HERE Art Center, New York (January).

——. (2003). *Future of Memory*. Set by Joel Sherry, Lighting by Susan Hamburger, Costumes by Katrin Schnabl. Performed by Danielle Goldman, Dawn Stoppiello, Michou Szabo and Sandy Tillett. The Duke on 42nd Street, New York (February).

——. (2009). *Loop Diver* (in-progress). Created by Troika Ranch (Dawn Stoppiello and Mark Coniglio). Set by Colin Kilian, Lighting by David Torish, Dramaturgy by Peter C. von Salis. Performed by Hank Hanstad, JJ Kovacevich, Johanna Levy, Travis Steele Sisk, Dawn Stoppiello and Lucia Tong. Commissioned by the Lied Center for Performing Arts, Lincoln, Nebraska (Premiere – October 2009).

Part 4

Corporeal Intertextualities – Body/Text/Technologies

13

Speaking for Performance/Writing with the Voice

Fiona Templeton

1 Speech is not writing by the body: writing is signs of speech

Orpheus's decapitated head, torn from his body by the female followers of the cult of Dionysus (the dark, the senses, life and death), was the book. Follower of Apollo (the light, the eye, the mind), the techné of his thought could remain after him, speaking as it rolled. In being separated from his body, it became the made thing, the defiance of death. It is the memory of speech, or the forecast of its possibility. Speech that imitates writing is papery.[1]

Performance is in some ways the opposite of the book. The book is mobile. Performance is in place and time.

Conventionally, writing for performance is texted on the page to be recreated outside of that space, in time, spoken. Creating for speaking, through, from and for performance but without writing, is texted upon, with and for time, by the body. The body is always in time. Unlike the page, unlike the reading mind, speaking for performance can hardly avoid being simultaneous with its co-textual speeches. Its structures are those of the voice, and the play and clash of voices. Its orality, however, is today not pre-scriptural. It can be informed by the memory of writing, or perhaps of the forecast of its possibility.

This chapter will describe and discuss a practice I have used for creating text for performance without writing.[2] This is a subjective account.

2 Creating through speaking

I began speaking into a recording device as a way of bypassing the experience of living through the blanks, the as yet unfilled pages that are always part of the writing experience, and that are collapsed for the reader. The equivalent is a blank on the recording (possibly the on/off click of the pause button, otherwise the duration of the sound of breathing), but lived thinking in time rather than confronted as a visibly unchanging space, the mirror of non-production.[3]

The practice was a form of discovery before it was a method of generating 'text'. But, gradually, what I was discovering was the range of means within it that made it most suited to how I wanted to relate to language, to think in language, and to be in a process of decision making about language.

I had a history of improvisational methods of generating language in performance, as distinguished from character-based or situation-based speech, so this new approach was more about thinking through speech than about enactment of speech, though that did also come into play, as described below, in terms of multiple voice.

The first few tries represent a progression of experiments: from simply reporting the scenes before the inner eye, hypnogogic imagery, to finding included in the visual fields fragments of language, morphemes which in turn began trains of meaning, to hearing such fragments, to interrogating these, to fully fledged dialogue, to multiple voices.

Regarding the dialogic, an aspect of the method may be represented by the following: in the 1990s I had translated Kleist's 'On the gradual production of thought by speaking'. For Kleist the presence of an addressee is important:

> There lies a special source of inspiration for him who speaks, in a human countenance facing him; and a gaze that lets us know that it has already understood a half-expressed thought, often offers us the expression for the entire remaining half.

While the recording device may not offer us a return expression (and I've written elsewhere about the speaker/listener or performer/audience relation), it is a less far stretch of the imagination for me to address a possible presence by speaking than by writing. But it was in particular the title of his essay that interested me.

As 'voices' or trains of thought replied to or interrupted each other, I wanted to give as many as possible of them room, despite the singleness in time of my speaking voice. Distinguishing 'speakers' came later, in the editing process, sometimes to 'make' a sense, although sometimes this was clear.

And so this practice goes beyond a field recording of myself, to a dialogue with the situation of recording.

At the level of the texture of the language, both language and thought seemed to be inventing themselves and each other. Above all, the experience of the process remains one of choices. While sometimes known features of oral poems came naturally, like rhythmic patterning, I could also conversely choose, at each juncture, to make a different kind of choice to those I had made till then. Then there was a new patterning, of the new and the known. This is a process familiar in experimental improvised music.

In a process where what has been made so far is less readily checked than on the page, different syntaxes evolve as well as different leaps of thought. Memory is functioning differently. At times the process is even amnesiac,[4] allowing the next point to be fully present. But at others, the sentence, if it is a sentence, is woven out of the cadences and clauses of breath and thought.

I would distinguish this practice from notions of automatic writing. In the generated parts of the language, the absence of the visual page allowed a different concentration, lived in silence not in blank. I was involved in and aware of the intense mental activity of most of these silences – they are suspensions into silence, not falling silent as a disappearance or end to thought. Often the cadence would remain lifted, the effort is audible in the voice, a word continued several minutes later after multiple silent expansions sideways from the crossroads of possibilities. The trope may not be unfamiliar to any writer, but here the journey was not from my mind to its expression via my hands, but via the more intimate organ of my mouth.

I was involved in how expression had to occur through an articulation of the embodied voice, the lips, tongue, sound and intimate movement; how sometimes this would even linger in the lallation[5] of effort of mind, the mind's trying to meet directly what it holds in its eye or the form it perceives, and the expression not being the faithfully teleported recreation of that onto the other bank, but the product of that meeting. Such forms might also include themselves the meetings of forms and thoughts and the meetings of phonetic responses.

The lacunae could represent, not a specific absence or omission, but a moment of thought, either too quick to record, or passing through territory that might not have been explored as the straightest journey to the page.

In a less discursive piece of speaking I might have described there (there being a moment, one of those lacunae that I can only tell you happened in the brief paragraph blank before this sentence) some feature of the landscape that struck me as I made my specific way along. Not a chance feature necessarily, though it might be, but perhaps also a feature conjured by the road.

And, unlike on the page that these texts found their way to in the editing process, silence returns in performance. In performance, silence is both the most and the least sensual part of speaking. The last because silence may be flesh closing upon itself while thought takes place, takes its place, and the most because in silence, the voice of the rest of the body can turn up the volume.

3 Speaking in time, speaking against time

a dead man talking, a dead foot walking[6]

In rows, on shelves: voices. Or voices stilled between closed covers, to speak when the covers are opened, to speak in the head or to speak in the mouth in an extraordinary ventriloquism across time. *I am thinking that thought, that exact thought that was thought before.* Or at least the form of that thought, its content re-coloured by now and by who I am. A time travel specifically permitted through the removal of the audible voice.

Like skulls in a crypt.

The book was both sinister and mobile, to be feared (West, 1984). The cult of Orpheus was a cult in fact not of the lyric song but of communication with the dead, a shamanic cult.

In this respect I want to mention two poets with whom, in the work created by this form of speaking-as-writing, I register affinity.

Alice Notley may, for all I know, sit in front of a paper or electronic page to write such works as *The Descent of Alette* (1996), but I was startled when I first found that on the published page (as opposed to either the writing page or the spoken recording), our texts looked the same: a series of small fragments of language made discrete by quotation marks.[7] Notley says that her use of the quotation mark is to separate the units of language, whereas for me they were to denote units of citation, the fact that I felt I was reproducing received language rather than simply generating it through distanced rational thought.

In Notley's *In the Pines* (2007), the conversation takes place between corpses. In *The Descent of Alette*, the speaker, if there is one, discourses with, and/or listens to, her dead father. The title is reminiscent of the Ancient Mesopotamian poem *The Descent of Innana to Hell* (Anon.,

1997), a journey to the afterworld or underworld, where Innana goes to recover her dead brother/companion Dumuzi. Notley has survived the deaths of two poet husbands, Ted Berrigan and Douglas Oliver. I would not draw conclusions as to similarities between Notley's methodology and mine. My observations are particularly about the layering of subject within subject, and the connotations of descent through such layers.

On the cover of Hannah Weiner's book *Clairvoyant Journal* (1978), is a picture of the author, on whose forehead is written 'I SEE WORDS'. In the texts of that period, the typed text is retained in publication, as Weiner uses it to represent the interruption of one phrase or word by another, one voice by another, usually hers by others. The text is interspersed with direct speech: 'appease us'. 'GO, HANNAH', and so on. In her later *We Speak Silent* (1997), lines are prefaced, play-like, by what appear to be the names of speakers – fellow poets, Bob Dylan, a polar bear, her mother, the living and the dead. In both books, these speakers (or voices, or units of language, or linguistic/performative positions) address her, reflect on and instruct the writing, even appear occasionally as physical descriptions, as seen. Occupation of this space, these language/subject places, is fluid, these times simultaneous. Something similar can be said of the non-hierarchical juxtaposition of subject-matters, philosophy and the quotidian jostling for page-time. The work is grounded in the specificity of these voices, and her place in them. The work can be knowingly comic.[8] And in *Clairvoyant Journal*, her retorts speak from a writing self different to the writing, different to the written self, besides the rest. Her text is peopled, and the reader, to negotiate it, joins the throng. Weiner both sees and hears, not just in the mind's eye, or rather the mind's eye is active along with her bodily ones, and bodies and the world become pages. And the typographic surface of the book attempts to embody their differences – capitals, underlines, between the lines, erasures, handwriting, falling lines.

Although I am not directly talking about performance here – these last two writers are poets – they are far from the often assumed notion that poetry speaks from a unitary first-person subject position. Rather than expressing thought, the position here is that language is active thought.

I've mentioned Weiner and Notley's work as examples of the specific power of language to traverse the boundaries between not only selves, but between the living and the dead, a power I previously connected to the book (Orpheus went to hell and back). And yet both poets' work lodges in multiplicity of voice, a performativity of the book.

4 Medead

The work I first made through the practice described in Section 2 above is *The Medead*, both the epic of Medea's journey, and Me Dead, another descent, out of self.

My interest in the mythic in the first place was partly linguistic. In writing or in poetry I had wanted, like the painter Elstir in Proust, 'to recreate things by removing their names' (Proust, 1999). And since myth has come to us through language on a journey that has ended clothed in caricature, I wanted to strip and re-speak that language, to remove the garb of mystery (since it tends to be garb that mystery is wearing), and to speak and hear the strangeness and actualness of its body.

It was in fact specifically before the vastness of the material relevant to this myth that I abandoned the page. I needed to know differently, and to let that manifest differently.

Some of the text is in sentences, some in phrases, some in words, some in even smaller units. Some are great swathes. Some units are word-like, or amalgamations of words, some can barely claim to be phonemes. They are not neologisms that I expect or would like to become words, rather I kept some of these un-English passages because I felt that they are activities present in language, they are time present in language, they are the tips or chips of icebergs, the mind (as mine was) coming to speaking through movements of the vocal body.

It would be inappropriate to explicate my own work here, and it would not of course be possible to reproduce my work by using the practice of speaking into a tape recorder – the practice does not explain the work. Like any writing practice, my ability to make choices and the choices I make are informed, as an improvising musician distils into the present a history of practice and memory, and the future path of the experiment.

But to give technical examples forged by the process, in words like 'marrangement', 'interprenjoyment', the amalgams emerged from attempts to let two verbal directions take place simultaneously in the mouth.

The separation into speakers registered various main modes: figures in the narrative, the narrative itself, and commentary. On the page these divide as text, chorus and birds. Simultaneity in language was something I had for some time been striving for, as the voices in, say, a Mozart opera (or more, how Busoni seems to play two or more different works on the piano at the same time), but being able to write thinking all of them, not writing them as separate. This emerged naturally from the practice of speaking:

Theseus: trickles her
Medea: lap drinks
Aegeus: just one
Chorus: *leans over*
Medea: it's closer than you think
Chorus: *she leans over*
Medea: this was where to stop
and me in it
the black sea shits out of his
Chorus: *so from his mouth*
scopes **Birds:**
Chorus: *different*
grunted
Medea: voice
Theseus: meat what I am

The birds are separated as they treat their lines differently, can repeat, interrupt, underscoreand so on. The birds are not the only place of lallation, and in fact they are often very coherent, like the birds of Scottish Mouth Music (Lomax, 1951). They might say:

know he needs you he needs you he needs you
know he needs you he needs you he needs you

and a person might say:

guide pal
body
otis
larming

curse

try
passage
war
mer
murse

There were occasions when I would come across a word or an idea when reading further relevant research, or I chanced to hear a word that I remembered having spoken in recording, though I hadn't known its meaning or sometimes even that it was a word or quite what it referred to. I would look it up, sometimes in an English, sometimes in a Greek dictionary, to find that it was absolutely relevant and exact. I don't think this was a mysterious process, but it certainly wasn't a fully conscious one. I had absorbed a quantity of material, but it was only through speaking that I had been able to bring it to the surface at the contextually accurate point.

The listening-back, then, was another stage, a hearing. Sometimes I wondered what I had whispered, at other times had to whisper it again to know, or to know why.

Though it is my means, I also don't privilege here the speaking voice. Sign-language, for example, uses a system that, unlike our alphabet, which purely represents sounds, includes the representation of whole ideas. Aaron Williamson's work involves the whole body in what often might also be called a linguistic enterprise in performance. He assumes the fact of his deafness as a form of knowledge. His book, *Hearing Things* (2001), is a transcription of performances originally recorded by speech-recognition software, although no recognizable speech occurred at the time, rather speech was created by the software in recognition of, for example, spluttering wax or moving furniture.

Though I describe in this chapter a practice of creating work that is later edited and expanded into the whole body and multiple persons, I've also used a similar method to improvise in actual performance. There I find my whole body does become involved, either in action begun as thought-impulses, or simply in the tensions, speeds, suspensions and turns of the movements of voice, mouth and thought.

I use a version of this too as the basis of an approach with actors to finding all that is in a text so generated, not to elide it into a norm of what it seems to say. In that process, the element of time reinserts itself inside the words, the voices reclaim their separate spaces there, and the body reflects the play of their simultaneous differences.

But in performance, not that it has less patience for the silences, but revealing them and living them would in fact give the lie to the collapsing of the time of thinking that the suspension in a word or thought assumes, is working towards, in the recording. Unidirectional time is not in any case what is going on.

To prepare for my next performance, I'm learning ventriloquism. I'm also beginning to see what happens if I use the time of the silences as a rhythmic fact, as physical as the impulses of tongue or air. Many of the effects of ventriloquism, of course, could be achieved with technology in the finished product. But I'm for now less interested in audio recording as a substitute for the human voice than in the process of dialogue between them.

Notes

1 In Dutch an over-literary playtext is *papierig*, literally *papery*.
2 I am distinguishing between performativity and embodiment, and also between the privately and publicly enacted.
3 This characterization of a blank on the page is the opposite of the active blank space in, for example, Larry Eigner's or Anne-Marie Albiach's writing.
4 I had begun an interest in this phenomenon in writing in *Cells of Release*, an installation of writing on a continuous strip of paper, because the spatial linearity of the work did not allow the cross-checking made easy by the back-and-forth weaving of lines on a page.
5 I use the term *lallation* for syllabic repetition, as in infant utterances sometimes thought of as pre-linguistic, or as a form of vocal musical production: 'lallation [n. of action f. L. lallare to "sing lalla or lullaby" (Lewis & Sh.) F.] †childish utterance.' *The Oxford English Dictionary* (1999).
6 From J. Jesurun, *Philoktetes*, New York: PAJ Publications.
7 These quotation marks were a form I abandoned later when I stopped distinguishing between the sources of the language in favour of attending to and making choices because of what was happening within it – or maybe this would be better described as letting and making what was happening within it through a series of choices. And at the time not all of the text is in quotes, as not all was 'heard' or 'read' in the mind's ear and eye, but I gave equal value to the volitionally generated and the apparently received.
8 The comedy of this saying-and-not-saying brings up Weiner's contemporary and champion Charles Bernstein, master of the place of the ironic voice, who, when he writes the most apparently naïve of poetry, positions himself with the ultimate irony (I thank Drew Milne for pointing out that this is opposite or perhaps akin to the Scottish colloquial use of 'double irony', saying what you mean as if you didn't) in saying that he meant every word of it. This relation with the words that one says, as if not necessarily being the things that one is saying, is a theatrical position, and yet, one says them. This digression into irony is because irony is a voice. In some ways, returning to the image of the poet at her or (in this case, his) desk, one can imagine an almost theatrical activity on the part of the writer surveying over there on the page what he is enacting saying. Maybe the fact that the transparency of the written voice is at issue is the flip-side of the choice not to write but to speak.

References

Anon. (1971). The Descent of Innana to Hell, in *Poems of Heaven and Hell from Ancient Mesopotamia*. Trans. N. K. Sandars. London: Penguin Classics.

Kleist, H. V. (1995 [1805/6]). 'Über die allmählige Verfertigung der Gedanken beim Reden', from *Philosophische und Ästhetische Schriften*. Trans. F. Templeton. *Mainstream Journal*.

Lomax, A. (Recorded by) (1951). Annie Johnston, Bird Imitations. Online <www.ubu.com/ethno/soundings/celtic.html>, accessed 23 December 2008.
Notley, A. (1996). *The Descent of Alette*. New York: Penguin Poets.
——. (2007). *In the Pines*. New York: Penguin Poets.
Proust, M. (2002). *In the Shadow of Young Girls in Flower*. Trans. J. Grieve. London: Penguin.
Templeton, F. (1997). *Cells of Release*. New York: Roof Books.
——. (n.d.). *The Medead*. The complete work is currently unpublished and unproduced.
Weiner, H. (1978). *Clairvoyant Journal*. New York: Angel Hair.
——. (1997). *We Speak Silent*. New York: Roof Books.
West, M. L. (1984, reprint). *The Orphic Poems*. Oxford: Oxford University Press.
Williamson, A. (2001). *Hearing Things*. London: Bookworks.

14
Authenticity and Perception in the Making of *Utah Sunshine*: A Dance Theatre/Arts Film

Ruth Way and Russell Frampton

Figure 14.1 Production still, *Utah Sunshine* (2007). Landscape, watcher and radio

Introduction

As artists/scholars we are engaged in constructing working methodologies and frameworks which allow us to explore and interrogate the interface between filmic representations of the moving body, spatial narratives and the process of embedding multiple temporal locations. This chapter addresses how we resourced and shaped our performance materials in the making of our recent film project *Utah Sunshine*.[1]

The film is primarily a dance and visual arts film reflecting the current concerns and investigations inherent in our own practices, these being somatic movement and collage-based mixed-media painting.

Utah Sunshine, in its making and critical practice, attempts to blur the distinction between personal space, theatrical digitized environments and external locations. Our intention is to create work that realizes the sensorial potential in these constructed, layered and imaginative landscapes. The development within *Utah Sunshine* of a complex system of interrelated textual statements, which refer to a diversity of cultural, historical and perceptual states, allows the film to construct a fluidity and multiplicity of meanings whose readings are dependant on the awareness, emotionality and cultural positioning of the viewer. The aim was to create a transcendent experience where interpretation surpasses rationality but connects the viewer with a deeper level of experience. The poetic nature of the visual and musical scores allow these meanings to be evoked rather than described, manifesting themselves through a refining process of excision, through editing and the compression of corporeal gesture in the filming process. The influential film-maker Robert Bresson is cited by Gilbert Adair as saying that creation is:

> A process primarily of ... paring of cropping and cutting away. What is the point, he argued of complacently revealing everything? What is the point of showing the whole when the part is capable of investing the same image with an even more profound mystery and rigor?
>
> (2007)

Utah Sunshine reflects this idea within its visual structure, by alluding to events and conditions in the iconic representations of archetypal scenarios that partly unfold,[2] but continue to remain ambiguous, just out of reach. This liminal space is created by the tension between the desire to resolve meaning and the fluid perceptual potential generated by this intertextuality.[3]

Authenticity, archaeology and the autobiographical

The working title for this project was originally '*Points of Impact*', and at the outset there was no conscious intention of developing a narrative relating to American nuclear testing in the 1950s, or including autobiographical elements from an individual's experience. These points of impact initially referred to collisions or forces generated through physical, emotional and temporal surfaces in conflict. The development of a more specific strand grew directly from the construction of a creative space, or network of potential, which sought to legitimize a sense of embodied authenticity by establishing a perceptually receptive and inclusive framework.

Our processes never dictated the representation of one narrative but opened up the possibility for these bodies to be authentic through their presence and their own embodied history in these constructed environments. The merging of computer manipulated imagery and the editing process developed this principle and provided further opportunities to express these multiple identities and stories. The direct impact of this nuclear testing on the human body and the environment, its consequences, history and associated iconography began to take on a particular resonance within the context of this experimental practice.

Michael Shanks in *Theatre/Archaeology*, where connections between performance and archaeology are proposed, draws our attention to 'the layering and the authenticity of depth, digging deep, looking to the significant detail, reading signs in the traces of things that have gone before' (Pearson and Shanks, 2001: 10). In *Utah Sunshine* this authenticity was grounded by Sondra Fraleigh's personal history and experience. In fact it was at a conference in 2002 that the impact of her account of how she and her family had been affected by the testing of nuclear bombs in Nevada left such a deep impression on me: 'As it turned out we were the fifty thousand expendables of the southern Utah towns, and as we had later learned, the atomic fallout had travelled on the wind at least as far north as Salt Lake City, they said the white ash that fell on us was inconvenient' (Fraleigh, 2004: 171).

According to Ann Cooper Albright:

> Autobiography, like dance, is situated at the intersection of bodily experience and cultural representation. Meaning literally 'to write one's life,' autobiography draws from its inspiration from one's being-in-the-world – that complex and often contradictory interaction of individual perspective and cultural representation.
>
> (Albright, 1997: 119)

The presence of Fraleigh's own reflective and impassioned voice leaving traces of these real stories became key to constructing the framework of the film and striving for a level of authenticity. In consultation with Fraleigh we compiled a list of ideas for footage, locations and movement possibilities for her to experiment with in Utah. Fraleigh's physical presence in the film reveals her embodied knowledge about these events and their consequences, allowing her to 'inscribe' her emotional authority and empathic understanding with the land in Utah and its ingrained personal histories.

Fraleigh walks through a field of grass swaying in the wind and enters an isolated cabin, a women lies there in a vessel, womb-like, as if in a cauldron,[4] *she sounds the structure's extended wooden feathers as if to recall these lost memories, Fraleigh's emotional presence appears as a ghosted figure in the doorway, her hand reaching towards her face calling reference to a sense of loss and remembered pain.*

The sentient body

In the rehearsal process one could say the dancers in *Utah Sunshine* were asked to surrender any notions of applying technical dance form to these fluid and constructed environments. There was never a directive to dance 'in time' or to follow choreographic material, but rather to be attentive and responsive to the different layering of time, space, image, rhythm and sound occurring at any given moment. In this sense we were weaving our own corporealities and physical energies with these texts and through the editing process, being visually stretched across these temporalities. I raise this point in connection with Lepecki's observations of 'the emergence of that new technique for disciplining the embodying of temporality: choreography' (Lepecki, 2004a: 126) and 'the deep complicity of dancing with the keeping of time' (ibid.: 126). Here Lepecki is drawing our attention to assumptions about dance's ontology and how this has been challenged, particularly by the recent 'conceptual or "minimal" contemporary dance scene in Europe' (ibid.: 127).

What interests me about this is how both the dance and the dancer can be released from this mode of being and operation. This became important for our own practice and constructing a relationship between a body politic and the political references and undertones present in the spoken and digitally manipulated text. These were also strategies to avoid being merely representative and didactic. When the dancers rhythmically twisted to *Let's Twist Again* (Mann and Checker, 1961), the focus was not on the development of synchronized movement material, but on a controlled rhythmic interplay between these repetitive and rigid movements with the looping projected backdrop. The aspect of 'control' has an implication here as this interplay attempted to expose both a social and political complicity with this period of testing and the subsequent sense of Cold War political paranoia.

Connections between past, present and future were also articulated through the bodily presence, as were differences arising from the three female performers, each coming from diverse movement backgrounds and experiences. Through the processes of image manipulation and editing, we became acutely aware of enabling each body to speak more authentically through their embodied practice and somatic movement responses. It was important to acknowledge these differences and work with them accordingly so as not to undermine the sentient presence of each body. The capacity to initially find a connection with one's own kinaesthetic experience in these charged spaces then offered the possibility of 'being moved' by them and finding an outer connection.

Informed by somatic movement practice[5] there was a conscious attempt to resist any fixing of the body but to facilitate the dancer '"to allow" rather than "to make" the movement happen' (Fraleigh, 2004: 169). These strategies encouraged us as performers to remain open to these fluid spaces and their semiotic and phenomenal potential, whilst at the same time having an awareness that 'objects of our perception are culturally and historically influenced' (Auslander, 2006: 194). This level of awareness has relevance in terms of our improvisational practice and the ability in one sense to work with a beginner's body, a body that becomes aware of its own patterning and cultural inscriptions.

The dancers 'just standing' in these spaces (Figure 14.1) and observing these events became a recurrent mode of being and a device to blur the distinctions between the performers' roles of witnessing, participating and being affected by the nuclear testing. Similarly, where the improvisation within the constructed sets had initially explored many different kinetic possibilities, the editing focused on poetic stillness and the physical contact with surfaces more than actual movement. The desire to charge each frame with an emotional resonance and for these corporeal bodies to express a connection with these landscapes began to take precedence.

The performers' behaviour in this work could be seen as being aligned to Victor Turner's understanding of liminal entities, which he describes as being, 'neither here nor there; they are betwixt and between the positions assigned and arrayed by law, custom, convention, and ceremonial' (1969: 95). The performers in *Utah Sunshine* exist on a threshold,[6] which we consider a creative space, allowing the breakdown of preconceived artistic/filmic procedures. The body, through its presence and its implicit, embodied narrative, bridges the space between the technological and the complexities of real stories, real people and real events. The performers' bodies metaphorically become this threshold, enabling their passage between multiple roles, environments and temporalities. This seamless, shape-shifting paradox reminds us of Lepecki's observation on the body in contemporary European dance, where he states, 'there is also an emphasis on the body in itself, in its bareness, in its superficial strength, in its massively complicated presence (2004b: 179*)*. In *Utah Sunshine* this complexity exits in the layering of the corporeal body positioned and stretched across and 'through' these textual interfaces.

The construction of a set of three slopes provided both a new environment and level for the dancers to explore. This played an important role in developing other ways of seeing and experiencing these landscapes, necessitating the different use of spatial orientations and kinetic relationships explored by both dancer and camera.

This spatial arrangement facilitated the layering of projected imagery behind the set and on the slope itself. One example of the integration of archive film and live movement which illustrates both a temporal and choreographic synchronization is a section of the film where we

Figure 14.2 Production still, *Utah Sunshine* (2007). Procession of souls, watcher

see archive footage of a line of girls walking in step across the desert (Figure 14.2). The 'live' figure leading the procession catches our eye as she interrupts the walkers' rhythm with a distinctive and quirky kick, which is echoed by the following 'archive' figure. This enigmatic processional walk draws our attention to a metaphorical and generational journey through this landscape, where each step seems to affirm our concrete relationship with the land and those who have passed through it before.

The compositional framing of the movement and gestures became refined by the parameters of this surface, 'movement becomes subordinate and intrinsically linked to the environment which contextualizes it' (Colberg, 1996: 45). This served to generate particular tensions as the dancers negotiated the gravitational forces encountered on this gradient. Movement exploration focused on the acts of falling, sliding, slipping away and conversely resisting these by physically interlocking to support one another until falling became inevitable. These physical restrictions further embedded a sense of authenticity through the body's repositioning and subsequent search for balance.

Landscape and the painter's perception

As a painter one aspect of my practice involves an articulation of the expression of landscape, not through the analytical and illusional process of actual depiction, but rather to give the sense of a perceptual experience itself through the same sensual processes with which we understand the complexity of the subject matter. This involves the construction of a lattice of meaning, a layered and inter-referential web which attempts to conjure up and record that moment where the landscape is perceived in its entirety, where perception and reflection join to produce Art. As Merleau-Ponty notes: 'The artwork ... unlike ordinary language and abstract thought, has a sensuous immediacy that comes close to our fundamental perceptual contact with the world. Unlike perception itself, however, it preserves and articulates the most crucial "invisible" scaffolding of the specific situation it is expressing' (in Crowther, 1993: 51).

The landscape itself is a hugely rich receptacle of meaning and textual potential, it is the physical space we inhabit and our relationship to it is fundamental. In many ways we are evolutionarily tailored to respond to this space, indeed at the deepest levels of our psyche, our very survival depends on an awareness and ability to interact with, and understand, its subtleties.

My personal painterly responses have lead to the development within my work of a network of interrelated visual components, derived from my contact with the external environment, where the act of looking becomes elevated to a process of absorption and the nuance and the fleeting become forces within the work which resurface, sometimes years later. The resultant paintings refer to a diversity of textual elements drawn from archaeology, history, climate, geology, personal and folk memory and myth as well as more formal considerations such as colour, composition, tone, texture, light and surface. These texts are then merged, juxtaposed and overlaid, edited in similar fashion to filmic processes, with resolutions dependant on the implied narratives and their relationship within the perceptual field that is the final painting.

Digital landscapes

The integration of technology is central within our working practice and aspects of digital technology were embedded throughout the project. These processes enabled us to charge a space with an atmosphere through

Figure 14.3 Production still, *Utah Sunshine* (2007). Narrator, figures and screens

sequences of layered projections, that at once both contextualizes the space and provides a creative backdrop facilitating physical responses to these texts.

A series of large screens were constructed for the purpose of projecting archive film material from US 1950s public information films of nuclear testing. Much of this archive material was then edited and manipulated, sections were looped and layered and through movement improvisational frameworks the dancers interacted with this digital landscape in a variety of spatial and empathic configurations. One example being the juxtaposition of scale, the dominant three-meter high projected image of the 'broadcaster's' face interrupted the normal sense of perspective, with the two performers occupying both extremes of foreground and background (Figure 14.3).

The sequence was extended by the addition of film of nuclear blast effects with an emphasis on the rhythmic pulsing of the forces impacting on the landscape. The movement improvisation was informed by a sense of journeying through this landscape, as if on a fixed boundary or border, building a tension between the live and virtual elements. This culminated with the dancers spinning 'dervish-like' in response to the climactic urgency implied by the footage, these trance like revolutions acted as a device to access another state of awareness and remembered experience and provide an entry point for a new scene.

This facilitation of technology enables a process of reductionism to occur, through the excision of material that does not directly 'speak' or hold a resonant meaning and the fine-tuning and intermeshing of sound and visual texts. Thus by striving to intensify each filmic sequence we attempt to distil something of the true substance of the encounter to the point where it becomes elevated into a fundamental perceptual experience, mediated through the body. As Merleau-Ponty states: 'I perceive in a total way with my whole being: I grasp a unique structure of a thing, a unique way of being, which speaks to all my senses at once' (1964: 50).

Textual surfaces

De Marinis's notion of intertextuality acknowledges the presence of those texts which have been previously explored. In *Utah Sunshine*, the development and sourcing of materials generated through our collaborative and experimental practice are complex and refer to our own individual disciplines and creative processes. This unfolding textuality is located within a wider ranging cultural sphere and relevant historical contexts, *Utah Sunshine* seeks to create a sense of an unending process

Figure 14.4 Production still, *Utah Sunshine* (2007). Slopes, screens, projected footage

of communication, where 'references and archetypes multiply to a point where they begin to talk amongst themselves' (Eco, 1986, cited in Lansdale, 1999: 15) According to de Marinis:

> It is the largely deliberate positioning of a creative work at the center of a rich network of echoes and references to other works ... the text ceases to present itself as a 'closed' entity and reveals itself instead as an unending process of production.
>
> (1993: 81)

The textual materials and cultural sources, which both inform our critical process and enliven our aesthetic eye are combined to produce surfaces of possibility. References which can appear obtuse and unexpected surface to contribute to a complex web of resonance.

In *Utah Sunshine* these 'echoes and references' included: John Ford's stylistic approach, specifically his visually influential westerns such as *The Searchers* (1956) with their focus on the primacy of landscape with characters playing out dramas that appear almost inconsequential against the grandeur of the terrain and the scale of geologic time. The road movie genre was also referenced, where the experience of landscape unfolds by one's passage through it, as was the notion of the outlaw or anti-hero as embodying qualities of the outsider or observer, removed from society but casting a chill glamour over proceedings. In the case of *Utah Sunshine* the cabin where Butch Cassidy was born in St George, Utah, became a location for sourcing film, and an element of charged historical authenticity was acquired. Other prime sources of material and references included vernacular dance of the post-war years, US public information films such as *White Sands* (1938), US Department of Interior movie newsreels such as *Atom Bomb* (1955), Joe Bonica and Sondra Fraleigh's somatic dance practice and texts, and personal accounts and memories of the nuclear testing in Utah.

The construction of a soundtrack that could reflect the complexity of the narrative strands and enhance the sensual experience of the body in these filmic landscapes was crucial. The diverse range of elements within the soundtrack included; the friction and impact of the body moving within the constructed sets; a series of archive sound recordings; the poetic readings; original music produced for the film, licensed music and ambient sounds.

As with the layering of the moving image within the film, the soundtrack functioned as a device to trigger archetypal responses in the viewer and create a network of cultural and historical reference points. The use of the well-known song *Let's twist again* acted as a catalyst to evoke nostalgia and memory with its recognizable beat and familiar lyrics transposed to accompany a series of images of nuclear destruction. This chilling juxtaposition heightens the sense of disorientation during this sequence, where archive footage of 'The Twist' and the live performers 'twisting' in front of the projections is intercut with rhythmically synchronized images of the apparatus of war. Robert Bresson's insight on the importance of rhythms and how these can penetrate, flood the senses and give shape to feelings identifies an aspect of our own approach to capturing the expressive body in our film making. 'Nothing is durable but what is caught up in rhythms. Bend content to form and sense to rhythms' (Bresson, 1975: 58).

The spoken narrative, which is both ambiguous and poetic is dispersed throughout our sourced media, enabling an intricate interweaving with the other sound elements. This was important in ensuring that words would not fix meaning or claim precedence over the presence of the body. The narrative moves between radio broadcasts, spoken text from archive film and Sondra Fraleigh's spoken readings, underlining the impression that this implied storyline is scattered through both time and space. The radio broadcast offers a powerful juxtaposition between the mundane and the poetic, injecting a political subtext rallying America against the 'corruptive and sinister' forces of communism with propagandist lines such as: '*America, with its four score beautiful stores and sparkling assortment*', and '*Who can help but contrast the beautiful, the practical settings of the Arcadia shopping hub and Woodyear quad with what you'd find under communism*' (*Utah Sunshine*, 2007).

Ephemeral moments such as Sondra Fraleigh's singular sharp inhalation underline the immediacy of experience, as it reminds the viewer of our corporeal needs and how life is sustained. This could refer to the first and last breaths we take, and perhaps defining the passage of a life. Subliminal, barely audible sonic undertones were placed beneath certain sections of the film, these included the sound of crackling embers and blast sounds from a nuclear explosion. This minimal layer of sound serves to further affirm the authenticity of the event, and builds a sensory, felt connection.

Conclusion

One of the most pertinent outcomes of this project has been the development of a sense of equality between the body and what could be considered to be the construction of a 'corporeal space'. This space enters a dialogue with the body, emphasizing its liveness and corporeal needs and also its capacity to remember lived experience. This is a space which is at once both sympathetic and empathic and one which nurtures the expressivity of the body. It is the extension of the corporeal body into its immediate charged environment where technology, the contextualized set and sonic soundscape coalesce. The authenticity in *Utah Sunshine* permeates these textual interfaces in an attempt to remind us of the connection between the land, its history and our body.

Notes

1 *Utah Sunshine*, 2007. A collaborative visual arts/dance film produced and directed by Russell Frampton and Ruth Way. Running time 14 minutes. Performers Ruth Way, Kristin McGuire, Sondra Fraleigh. Music: Ben Davis [Basquiat Strings], Max de Wardener. Shot on Mini DV in the United Kingdom, France and Utah, USA.

2 An Archetype, in this sense referring to the notion of a collectively inherited unconscious idea, an image or perceived distillation of experience into a symbolic/iconic form which exists universally in the individual psyche.

3 In *Dancing Texts* Janet Adshead Lansdale refers to Kristeva's notion of intertextuality as a ' mosaic of quotations' (Kristeva, qtd in Moi, 1986: 36) and Kristeva's argument that the interpretative process is the 'creation of a dialogue from an intersection of textual surfaces' (Adshead Lansdale, 1999: 15).
4 The mythological cauldron appears in many diverse cultures and is usually connected to the idea of a representation of a gateway to the underworld. The cauldron can be seen as a receptacle of wisdom and knowledge perhaps symbolic of the mother's womb and a location of rituals of initiation.
5 Somatic practice described as: 'A desire to regain an intimate connection with bodily processes: breath, movement impulses, balance and sensibility ... a much broader movement of resistance to the West's long history of denigrating the value of the human body and the natural environment' (Johnson, 1995: xvi).
6 A liminal threshold as defined by van Gennep and Turner to mean 'a transition between' (Turner, 1982: 41), a state of ambiguity, of dissolving of identity and disorientation, ritualistically linked to ideas of rites of passage.

References

Adair, G . (2007). 'The supreme genius of cinema', *The Guardian*, 10 October. <http://film.guardian.co.uk/features/featurepages/0,,2187659,00.html>, accessed March 2008.

Albright, Ann Cooper. (1997). *Choreographying Difference: The Body and Identity in Contemporary Dance*. Hanover, NH: University Press of New England [for] Wesleyan University Press.

Auslander, Philip. (2006). *'Afterword:* Is There Life after Liveness?', in *Performance and Technolog*. Ed. S. Broadhurst and J. Machon. Basingstoke: Palgrave Macmillan.

Bresson, B. (1986). *Notes on The Cinematographer*. Trans. Jonathan Griffin. London: Quartet Books.

Colberg, Anna Sanchez. (1996). 'Altered States and Subliminal Spaces: Charting the Road towards a Physical Theatre', *Performance Research*, 1.2 (Summer): 40–56.

Crowther, Paul.(1993). *Critical Aesthetics & Postmodernism*. Oxford: Clarendon/ Oxford University Press.

Eco, Umberto. (1988 [1986]). 'Casablance: Cult movies and intertextual collage', in *Faith in Fakes*. Ed. D. Lodge. London: Secker & Warburg.

Fraleigh, Sondra. (2004). *Dancing Identity, Metaphysics in Motion*. Pittsburgh: University of Pittsburgh Press.

Johnson, D. H. (ed.). (1995). *Bone, Breath, and Gesture – Practices of Embodiment*. Berkley, CA: North Atlantic Books.

Lansdale, Janet Adshead. (1999). *Dancing Texts: Intertextuality in Interpretation*. Alton: Dance Books.

Lepecki, Andre. (2004a). 'Exhausting Dance', in *Live Art and Performance*. Ed. A. Heathfield. London: Tate Publishing.

——. (2004b). 'Concept and Presence in The Contemporary European Dance Scene', in *Rethinking Dance History*. Ed. A. Carter. London: Routledge.

Marinis, M.de. (1993). *The Semiotics of Performance*. Bloomington: Indiana University Press.

Merleau-Ponty, Maurice. (1964). *Sense and Nonsense*. Trans. H. Dreyfus and P. A. Dreyfus. Evanston, IL: Northwestern University Press.

15
(Syn)aesthetic Writings: Caryl Churchill's Sensual Textualities and the Rebirth of Text

Josephine Machon

> The transfinite in language, as what is 'beyond the sentence', is probably foremost a going through and beyond the naming. This means that it is going through and beyond the sign, the phrase, and linguistic finitude.
>
> (Kristeva, 1992: 190)

> We have to discover a language which does not replace the bodily encounter ... but which can go along with it, words which do not bar the corporeal, but which speak the corporeal.
>
> (Irigaray, 1985: 43)

Current practice has proven that performance continues to reinvent itself in this highly technological age. This is not only true of live art, contemporary dance and devised practice but it is also the case for playwriting. Since the 1990s, an approach to writing for performance has emerged which fuses the visceral and the technological in the very fibres of the text.[1] Such playwriting has caused a paradigm shift in the process, forms and content of textual practice. It has served to reformulate live performance in the 'traditional' theatre-writing arena, cross-fertilizing the poetic and the political, the esoteric and the accessible. Most importantly, it continues to demand new strategies for performance and makes innovative interdisciplinary practice a prerequisite for production.

Caryl Churchill's work is exemplary of such textual practice. Her approach, which constantly reinvents the possibilities of theatre, has seen to the rebirth of writerly textualities in current theatre.[2] Hers is a practice of *play*writing produced from the 'play' with the complexities and possibilities offered in performance. This is apparent in Churchill's formalistic experimentation with unusual image and intense physicality, woven into the very fabric of the *play*text itself. This encourages practical inventiveness from directors, designers and performers alike which includes a return to the body as the primary locus of signification, as well as inviting imaginative technological experimentation to make manifest ideas and imagery embedded in these texts.

As this suggests, underpinning Churchill's writing is a return to the 'thinking body'; a penetrating understanding of the diverse ways in which the body both communicates *and interprets* in the live performance moment. In particular, she manipulates the human body as a sensual signifier and receiver of lived personal, social, political and historical experience. The body of performer and audience member alike becomes the sensate conduit for communicating and receiving her narrative, thematic and formalistic concerns.

Applying (syn)aesthetics (see Machon, 2009) as the defining style and strategy of appreciation of such work, this chapter will examine the experiential quality of these sensual textualities in this technological age.[3]

Sensual textualities: interdisciplinary *play*writing

> We write in confrontation, through the love of hand-to-hand fighting with our language. ... My language is not in my mind, like a tool that I would borrow in order to think. It is entirely within me: words are our true flesh.
>
> (Novarina, 1996: 125)

> The triumph of pure *mise-en-scène*.
>
> (Antonin Artaud, qtd in Derrida, 1978: 236)

As illustrated by Churchill's work, the (syn)aesthetic performance style explores various combinations of verbal, physical, design and technological texts within a (syn)aesthetic hybrid. A particular predominance is given to playfully disturbing written texts, referred to here as *play*texts, that are marked by a visceral-verbal quality.[4] With (syn)aesthetic work, subsequent processes of intellectual analysis rely on an individual's corporeal memory of the piece. In embracing intertextual practice, the (syn)aesthetic style celebrates the play between, and flux within, the linguistic, corporeal, visual, aural and technological. (Syn)aesthetic writing overturns traditional notions of textual practice in theatre and allows for a rebirth of what textual practice can be in performance. Such *play*writing can juxtapose a variety of linguistic registers as well as emphasizing the corporeal and interdisciplinary within its very form.

Churchill's focus on the human body and its potential to make and undo meaning in diverse ways can be charted from *Vinegar Tom, Cloud 9, Fen* and *A Mouthful of Birds* (see Churchill, 1996, 1997c; Churchill and Lan, 1998), to the present.[5] Furthermore, her writing continuously crosses boundaries and cross-fertilizes itself with other disciplines *in order to produce* a visceral impact. Churchill interweaves diverse linguistic styles with dance, music and design; elements written into the very substance of the *play*text. This is true of the hybridized *The Skriker* (Churchill, 1994) and *Lives of the Great Poisoners* (Churchill, 1998), to her verbal soundscores for dance, *Fugue* and *Hotel* (Churchill, 1998, 1997b), to the *play*text within the Siobhan Davies Dance Company's installation, *Plants and Ghosts* (Churchill, 2002b).

With Siobhan Davies's Company's *Plants and Ghosts*, Churchill's verbal soundscore playfully expounds the notion of evolution, where the details of the development and breakdown of a relationship is drawn out from a single sentence: 'She bit her lip.' This voiceover starting point (performed by Linda Bassett), packed full of emotion, potential and foreboding, evolves into a visceral-verbal soundscore which dances through repeated phrases. Yet this is a repetition that is made different due to the extension of each line to elucidate the fully formed situation of a painfully disintegrating relationship. Following the rhythms, twists and extensions of the verbal phrasing the dance vocabulary is built up out of multiplied gestures, eloquently and progressively more urgently signed by a lone female dancer; the word-movement exchange sensually fusing the thematic concept with the evolving narrative, fusing both across dancer and audience. This sequence in turn fusing with the evolving digital soundscores and choreographies either side of the linguistic/physical play. As this suggests, with Churchill's texts, the physicality in the visceral-verbal text demands to be interpreted through striking physical imagery and invites accompanying digital play.

Throughout her practice Churchill has persistently played with different dimensions, times, worlds, demanding that the audience engage with realms that activate the imagination and disturb human cognition. This ludic subversion can be traced from early plays such as *Not ... not ... not ... not enough oxygen*, through *Cloud 9* and *Fen* (see Churchill 1993, 1997c, 1996) to *Far Away, A Number* and *Drunk Enough to Say I Love You?* (2000a, 2002a, 2006a). The recent 2008 celebration of her work in a series of rehearsed readings at The Royal Court Theatre in London is testament to this. Within all her work this 'realm-play' serves to upset logic and disturb conventional capacities for meaning-making within performance. This otherworldly play of image and action alone opens itself up to exciting possibilities in technological interpretation.[6]

The realms that Churchill presents on stage play with liminal space, explore the in betweens of experience; between life and death; between reality and fantasy; between past and present, present and future; between logic and the illogical; between madness and sanity; between femininity and masculinity, between the tangible and intangible. She opens up the performance space in time and location to allow the unthinkable to happen, truly taking the audience to what Susan Broadhurst refers to as 'the edge of the possible' (1999a: 1). In much of Churchill's work there is an all-pervasive sense of the nightmare, of hauntings, of intoxication, of transgression and disturbance in thematic and formalistic concerns. Such hauntings remain as traces within the corporeal memory of the audience's experience of the ideas, images and narratives received.[7] Once again, such play with(in) and between time, space and states encourages experimentation with form that may be executed via the traditional age-old technologies of body, space and sound. Alternatively, they are open to the exciting play of realm offered by digital and virtual worlds.

In probing the possibilities of the imagination and theatre, Churchill's writing exploits the potential for a variety of performance elements to force each other into new dimensions for communication. As a result, Churchill steadfastly deconstructs boundaries, shifts and disturbs the divisions of performance conventions, demanding that these be constantly destroyed and reinvented. Her intertextualization of dance, music and design within the substance of the text, sometimes explicit as with *The Skriker*, often implicit as with the potential for technological play throughout *Far Away* or *Drunk Enough to Say I Love You?*, demonstrates that *play*writing can be perceived as a physicalized and liminal practice in itself, with an indefinable nature and inherent resistance strategies.

Churchill's recent work such as *Far Away*, *A Number* and *Drunk Enough to Say I Love You?* are intensely concentrated full-scale *play*texts that crystallize ideas and are tantalizingly open in form, requiring imaginative realization in production. Consequently, the smallest details of space, design and movement, like the structure of the *play*texts themselves, *feel* vast yet (much like Samuel Beckett's works) are often minimalist in execution.

With the original production of *Drunk Enough To Say I Love You?* (2006b), the final sequence, slowly, subtly, sees the sofa (in this production, the central focus of the interplay between Jack and Sam) rise to the rhythm of the final speech. Surrounded by darkness that magically, ridiculously, conjures and swallows up mundane artefacts, such as coffee cups or cigarettes, the sofa imperceptibly moves higher with each scene. This communicates the idea that these men, these countries, their ongoing personal/Political relationships, are moving farther away from solid ground. It reinforces the sense of Jack and Sam losing contact with reality via entirely haptic means.[8] We both see and *feel* them suspended in space, isolated in a world of their own creation. Like the dysfunctional intercut, unfinished rhythms of the speech, this final image leaves them (and the audience) hanging in the air, pondering their destructive, three-dimensional flight of fancy:

JACK: catastrophe
SAM: so fucking negative
JACK: frightened
SAM: leave me if you don't
JACK: done that
SAM: stay then and be some
JACK: hopeless
SAM: and try to smile
JACK: dead
SAM: because you have to love me
JACK: can't
SAM: love me love me, you have to love me, you
End.

(Churchill, 2006a: 42)

As this play with the rhythm of speech illustrates, Churchill's work breaks away from narrative, subverts dialogue and rewrites linguistic conventions in order to (re)present and make sense/*sense* not only of human relationships but also the social, cultural and political mood of the time.

Disturbing speech patterns

> We've got ninety-nine per cent the same genes as any other person. We've got ninety per cent the same as a chimpanzee. We've got thirty per cent the same as a lettuce. Does that cheer you up at all? I love that about the lettuce. It makes me feel I belong.
>
> (Churchill, 2002a: 50)

Churchill's writing highlights a linguistic return to visceral qualities of communication. This encompasses both the ability to stir innermost, inexpressible human emotion and to disturb those viscera that cause aural, visual, olfactory and haptic perception. In this way, her manipulation of speech takes on the double-edged quality of making-sense/*sense*-making akin to the (syn)aesthetic style. As *Far Away* illustrates, with (syn)aesthetic *play*texts the word is defamiliarized and has to be (re)cognized and made sense of via a sensate fusion of verbal and non-verbal means. Within a (syn)aesthetic appreciation process, a certain semanticizing of the somatic experience of words occurs during and/or following a performance, where the 'meaning' of the words is reflected in both sound and *feeling* (that is, emotion and hapticity) that they embody.

Churchill's writing has also explored the border between language and sound, demonstrating the effects of language at its most damaged and destroyed in order to reve(a)l it in its sensate and physical quality. Defamiliarized language, like that presented in Churchill's *The Skriker, Blue Heart, Far Away* or *Drunk Enough to Say I Love You?*, shows how the verbal can be replayed, destroyed and reinvented in order to produce a more visceral form of verbal communication and thereby find the somatic essence of words and speech. With *Far Away, A Number* or *Drunk Enough To say I Love You?*, it is the rhythms, tones and textures of what *is not* said, as much as the increasingly unusual manipulation of what is said, which harnesses the audience's imagination.

Here words and the carefully constructed space in between the words have the potential to transmit emotive and sensate experience and take on an Artaudian quality; a visceral-verbal action that lacerates. The spoken and unspoken etches itself into the perceptive faculties of the human body. This ludic disturbance of language can discomfort and unsettle the audience. It causes a recognition of speech forcing an equivalent recognition of ideas, events, states, experience and so on to achieve a new point of verbal making-sense/*sense*-making. Language in this mode becomes far more than merely aural description. It is able to penetrate deeper; word as a fusion of somatic sound capsule and disfigured semantic sign.

As a result, in Churchill's texts, speech interwoven with other components of the (syn)aesthetic hybrid becomes a key sensate component within its fused corporeal communication. As suggested earlier with the *Plants and Ghosts* sequence, the fusion of words with other elements of the hybrid translates visceral experience to the audience and enables that the suturing of the verbal and physical, the conceptual and experiential, is achieved. The predominance of the senses and corporeality in Churchill's writing ensures that it is a very real *writing* of the body in concept and form. To illustrate, in *The Skriker,* in the same way that the shape-shifting body of the Skriker is twisted and deformed when we first see it, Churchill twists and deforms meaning and confounds expectation in language, reve(al)ling in its rhythmical, kinetic potential to communicate;

> SKRIKER: Heard her boast beast a roast beef eater, daughter could spin span spick and spun the lowest form of wheat straw into gold, raw into roar, golden lion and lyonesse under the sea, dungeonesse under the castle for bad mad sad adders and takers away. Never marry a king size well beloved.
>
> (Churchill, 1994: 1)

Churchill takes this de(con)struction of language further in *Blue Heart* (1997a) where in the first part, *Heart's Desire* (Churchill, 1997a: 3–36), she upturns expectations in the form and content of the *play*text, and then in the second, *Blue Kettle* (Churchill, 1997a: 37–69), she pares language down to its most basic of sounds. In this way image and sound are semanticized; when performed, the 'meaning' of words are reflected in the sound and *feeling* they embody as much as in their semantic capability. This enables the individuals in the audience to perceive the details aurally, intellectually, corporeally. The work becomes both *felt* and *understood* in an entirely embodied way;

> DEREK: Ket ket still ... I'm still ket I am ... if bl like me.
> MRS PLANT: T t have a mother
> DEREK: K
>
> (Churchill, 1997a: 68)

Consequently, Churchill's *play*texts demonstrate an ongoing (syn)aesthetic renegotiation of linguistic and performance structures in order to disorientate and unsettle expectation and perception in the audience. Her playful disturbance of verbal acts becomes, simultaneously, a critique, a re-evaluation and a celebration of what a *play*text is and can be. Manipulated (syn)aesthetically in this way, Churchill's verbal texts evoke the ineffable, transcending speech as we 'know' it. *Far Away, A Number* and *Drunk Enough to Say I Love You?* are emblematic of Churchill's manipulation of language and of visual image through the most minimal means. It is this potent distillation of the speech and a focus on the imagination that leads to Churchill's idiosyncratic crystallization of form; her compulsion to delve deeper into the possibilities of human perception through the power of *play*text.

In terms of the unusual exchange between performance and audience, by 'recasting language' Churchill taps a pre-verbal consciousness which allows the 'instinctual body' to 'cipher' the words (Kristeva, 1982: 61), drawing on corporeal memory in interpretation. As a result, Churchill's practice, paradigmatic of (syn)aesthetic *play*writing, highlights the embodied exchange that occurs in visceral performance between writer, performer and audience member throughout the process of production, from the initial act of writing to the ongoing process of interpretation. Her approach focuses on the live moment of performance and interrogates the essence of the live theatrical event.

All of these examples of Churchill's unique experimentation with form are testament to the fact that, where traditional play-texts have previously been considered to be reductive, enforcing closure in meaning-making processes, with her (syn)aesthetic *play*texts she establishes an opening process in terms of immediate appreciation and subsequent analytical strategies. It is the particular manipulation of speech, gesture and image within her written texts, alongside the layers of sensual textualities within the (syn)aesthetic hybrid (bodies, light, sound, technologies and so on), that work with and/or against the words, which ensures a powerful (syn)aesthetic response in appreciation is achieved.

Writing and the (syn)aesthetic sense

> The theatre is ripe for crystallizing language.
>
> (Artaud, 1974: 179)

> [I]t is possible to present the 'unpresentable'... from beyond but also including language.
>
> (Broadhurst, 1999a: 8)

Churchill crystallizes and concentrates the intensity of personal, lived experience and themes, revealing the invisible (experiences, emotions, states, concepts) through word and visual image. Furthermore, Churchill's writing demonstrates how (syn)aesthetic *play*texts connect wider social, historical and cultural issues with the individual and personal in an unusual and evocative way. Churchill's *play*texts have the power to make tangible the intangible, where words have the ability to activate the '(syn)aesthetic-sense' and touch the unconscious.[9]

Broadhurst, following Jean-François Lyotard, highlights the experience of the unsayable as that 'something which should be put into phrases, cannot be phrased' (1999b: 21). With Churchill's (syn)aesthetically styled speech, this 'something' *has* been phrased in such an unusual and immediate linguistic manner, it foregrounds that which formally denied phrasing. The noetic capabilities of language in (syn)aesthetic *play*texts comes about because the audience hears the words first with their bodies, with a primordial sentience. Etched onto the bodies of the audience, the words themselves become corporeal citations in subsequent appreciation, which aids the experiential understanding of the concepts at the heart of the work.

This is particularly true of *Far Away* which plays with the imagination and concentrates language and image to an essential form. In doing so, Churchill engages human logic in an unusual way. Linda Bassett, who played Harper in the original Royal Court production (Churchill, 2000b), describes how in wrestling with the text as a performer the very form of the writing excites the brain. As a result this is relayed in the appreciation of the delivery of the language, where, 'like lateral thinking games' the visceral-verbal texts allow the mind to make 'fascinating journeys' where 'you can feel your nerves, your brain cells being forced into a different synapse' (Bassett, qtd in Machon, 2009: 151).

The penetrating verbal images and concentrated actions written into *Far Away* demand an equivalent response in the visual realization. This is exposed in particular in the dance-like hat-making sequence that occurs in the second act. A visual play with the design of the hats is accentuated in the details of the actions:

> *JOAN and TODD are sitting at a workbench. They have each just started making a hat ... . Next day. They are working on the hats, which are by now far more brightly decorated ... . Next day. They're working on the hats, which are getting very big and extravagant ... Next day. They are working on the hats, which are now enormous and preposterous.*
>
> (Churchill, 2000a: 16–22, emphasis original)

Not only is a subversive approach necessary in the design of the hats but the actions, alongside the rhythm and content of the dialogue that ensues, demands that an intensely physicalized creation of the hats is made clear. As a result, the hat sequence conveys a formal expression of the intensely felt relationship that is developing between Joan and Todd. This balletic sequence leads to the visually disturbing presence of the hat parade:

> *Next day. A procession of ragged, beaten, chained prisoners, each wearing a hat, on their way to execution. The finished hats are even more enormous and preposterous than in the previous scene.*
>
> (Churchill, 2000a: 24, emphasis original)

Accompanied by distorted, unsettling cavalcade music, the chilling movement of the hat parade produces a disquieting effect, which draws on traces and echoes of lived historical events (Churchill, 2000b). The prison clothes, the bodies chosen (all shapes, sizes and ages – most disturbing the fact that the wearer of Joan's winning hat is a child, clinging onto an adult's hand) instantly evokes the holocausts, genocide, of recent history – Auschwitz, Cambodia, former Yugoslavia, Rwanda, Iraq and on. Yet, the presence of the hats serves to push this image further, delving deeper into the dark and dangerous possibilities of human nature. It is the concentration of the images found in Churchill's words that engage a (syn)aesthetic-sense and provide an embodied understanding of a simultaneously abstract and primordial idea – the notion of the human heart of darkness.

The concentrated style of Churchill's extraordinary language further explores the noetic capabilities of the imagination in the final speech. Here the sensational and conceptual are fused; ideas are made tangible and emotional and sensational resonance is forced through the body with the visceral trigger of each and every word. The absurd, lyrical, cruelty of the piece is delivered with absolute solemnity, absolute brevity – all the more defamiliarized and disturbing for it. In its ludically disfigured language it explores the apocalyptic, irreparable damage that humans are capable of on an intimate and personal scale as well as in socio-political ways (to be developed later in *Drunk Enough To Say I Love You?* and *Seven Jewish Children: A Play for Gaza* (Churchill, 2006a, 2009)).

This final speech is exemplary of Churchill's ability to concentrate form and verbal play in order to expose themes and philosophical concepts through the distillation of what is said, which enables the audience to perceive the undercurrents of what has not been voiced; condensing time and articulating vast concerns. The references to the animals, vegetables and minerals, all forces of nature, that have taken arms against each other again draws on traces and associations of lived history, alongside the apocalyptic and destructive creation that initiated human evolution, pointing towards human culpability in climate change. Yet Churchill unsettles and subverts this by the description of the atrocities witnessed:

> [T] here was one killed by coffee or one killed by pins, they were killed by heroin, petrol, chainsaws, hairspray, bleach, foxgloves, the smell of smoke was where we were burning the grass that wouldn't serve. The Bolivians are working with gravity, that's a secret so as not to spread alarm. But we're getting further with noise and there's

> thousands dead of light in Madagascar. Who's going to mobilise darkness and silence?
>
> (Churchill, 2000a: 38)

As a result this speech enables an encounter with ideas as much as actual presence through the viscerality of the language and the ideas it suggests, which evoke a (syn)aesthetic-sense.[10]

With this speech, Churchill engages words in a manner 'distinct from their actual meaning and even running counter to that meaning' by way of creating 'an undercurrent of impressions, connections and affinities beneath language' (Artaud, 1993: 27). Here Churchill's experiential visceral-verbal gives 'words something of the significance they have in dreams' (Artaud, 1993: 72). Kathy Tozer, who played Older Joan in the original production of *Far Away*, refers to the 'transcendental' quality of the final speech (Tozer, 2001). Just as Older Joan closes the play with, 'I put one foot in the river ... when you've just stepped in you can't tell what's going to happen' (Churchill, 2000a: 38), via these defamiliarized words that cause a recognition in terms of meaning, the audience are enticed to place a foot in the water; metaphorically, throughout the play the audience are at the liminal point of a realization of the ideas communicated. In this original production the safety curtain crashed down at this point, like a physical representation of the forces of *being in the moment* of recognition. This provides a visual equivalent to the forceful experiential quality of the (syn)aesthetic-sense, when ideas and experiences are presented on a visceral level, as if for the first time.

In this way *Far Away*, in its intensely concentrated structure, is experiential with a viscerality that violates performance expectations in form and content, returning humans to the senses in a technological age. It is a prime example of how Churchill reawakens the possibilities of what writing is and can be in current and future practice.

The rebirth of text ...

> There is a nascence and a renaissance, an amorous interchange, and a perceptual resurgence within writing. Writing is resurrectional.
>
> (Novarina, 1996: 108–9)

Churchill's visceral-verbal style actuates a (syn)aesthetic-sense in the immediate experience of the work and the subsequent processes of recall. She creates texts that demand a visceral response in terms of their verbal, physical, spatial, visual and technological realization.

The audience is challenged to confront an intuitive response in order to appreciate her work, thereby prioritizing individual interpretation. Churchill's writing practice blends the aural, visual, olfactory, oral, haptic and tactile to create intersensual work. This activates a double-edged making-sense/*sense*-making process of appreciation and ensures that the performances of these *play*texts become an experience in the fullest sense of the word; to feel, suffer, undergo.

Churchill's *play*texts present a practice of *play*writing produced from the play with the complexities and possibilities offered in performance. They encourage inventive interpretation from any practitioner involved in their realization, often demanding a return to the human body as the primary locus of signification fused with imaginative technological experimentation to manifest the ideas and imagery embedded in these texts. Such (syn)aesthetic *play*writing inspires an embodied exchange between writer, performer and audience member. The focus on diverse approaches to the live moment embedded in such writing contravenes categorization and foregrounds the experiential. Consequently, Churchill's (syn)aesthetic writings reawaken the possibilities of what writing is and might be in traditional theatre environments and signal the rebirth of writerly textual practice in a technological age.

Notes

1 I use 'visceral' to denote those perceptual experiences that affect a very particular type of response where the innermost, often inexpressible, emotionally sentient feelings a human is capable of are actuated. Visceral also describes that which affects an upheaval of the physiological body, so literally a response through the human viscera.

2 'Writerly' follows the post-structuralist theories of, in particular, Roland Barthes, Hélène Cixous, Jacques Derrida, Luce Irigaray and Julia Kristeva. Raman Selden, Peter Widdowson and Peter Brooker define writerly texts as those that encourage the receiver 'to *produce*', and play with, 'meanings' rather than simply consuming a specific 'fixed' meaning (1997: 159, emphasis in original); style as much as content often being multilayered, shifting and ambiguous. Deconstruction is the tool employed to interpret such texts as definitive readings are proven to be impossible, or futile. As a result, there can only be free-play within interpretation as evidenced in the deconstructive approach akin to (syn)aesthetic analysis.

3 (Syn)aesthetics derives from 'synaesthesia' (the Greek *syn* meaning 'together' and *aisthesis*, meaning 'sensation' or 'perception'). Synaesthesia is also a medical term to define a neurological condition where a fusing of sensations occurs when one sense is stimulated, which automatically and simultaneously causes a stimulation in another of the senses. So in terms of this neurocognitive condition, synaesthesia is defined as the production of a sensation in one part of the body resulting from a stimulus applied to, or perceived by,

another part. Within this I also employ the definition of 'aesthetics' as the subjective creation, experience and criticism of artistic practice. Following all of these definitions, my intention is to fuse ideas held within the medical term with those surrounding the aesthetics of performance practice. My reworking of the term as '(syn)aesthetics', with a playful use of parenthesis, encompasses both a fused sensory perceptual experience and a fused and sensate approach to artistic practice and analysis. The parenthesis is also intended to distinguish this performance theory from the neurological condition from which it adopts certain features and to foreground various notions of slippage and fusing together in arts practice and analysis. It is a strategy of analysis that prioritizes individual, immediate and innate processes of recall and provides a mode of analysis for non-genre specific, visceral performance. (Syn)aesthetic analysis comes into play where the form and content of the artistic work is executed and received in a way that fuses the somatic ('affecting the body' or 'absorbed through the body') and the semantic (the 'mental reading' of signs) ensuring a double-edged making-*sense*/sense-making process is affected in interpretation. See Machon, 2009, regarding this process of analysis and for a detailed explication of the particular strategies integral to (syn)aesthetic performance practice.

4 *Play* is fundamental to the impulse in creation and appreciation of the (syn)aesthetic style in practice and analysis. This draws on Immanuel Kant's '*free play* of imagination' (1911: 58–60, emphasis in original); a 'pleasure that depends ... on consciousness of the harmony of the two cognitive powers imagination and understanding' (Broadhurst, 1999b: 28). Derrida asserts '[w]riting *represents* (in every sense of the word) enjoyment. It plays enjoyment, renders it present and absent. It is play' (1976: 312, emphasis in original).

5 As Elin Diamond argues, the body in Churchill's writing is 'a special site of inquiry and struggle' and her writing itself 'empowers speakers with vital words, incites bodies to move in space' (1997: 83).

6 I have experienced this repeatedly with student experimentation with Churchill's *play*texts, where undergraduates play with the possibilities for embodied visual realization of the ideas via physical imagery, film, soundscapes and digital interplay with live performance.

7 Geraldine Cousin suggests Churchill employs the 'stage as a place of magic possibilities' (1989: 61). Richard Eyre and Nicholas Wright highlight how this reveals 'new worlds beyond and beneath the surface of ordinary life' and exposes a 'secret underside: magical, sexual, criminal', where there is a 'shifting of the ground' beneath the feet of her characters' (2000: 294–5). According to Diamond, it is this play with 'representational space' which alters the audience's perspective on 'the play's "world"' (1997: 92) and thus their own worlds. Magic, play, transformation, transgression and disturbance, all work together for the releasing of 'alternative realities' and allows a 'movement away from constraint towards the freeing of possibilities' (Cousins, 1989: 61). Specific examples of those *play*texts that explore such realms are *Vinegar Tom, Fen* (see Churchill, 1997c, 1996), *The Skriker* (Churchill, 1994), *A Mouthful of Birds* (Churchill and Lan, 1998), *Far Away* (Churchill, 2000a), *A Number* (Churchill, 2002a) and *Drunk Enough To Say I love You?* (2006a).

8 I use the term 'haptic' (from the Greek, 'to lay hold of') alongside 'tactile' as the latter tends to connote only the superficial quality of touch. Haptic, taken from Paul Rodaway's usage, emphasises the tactile perceptual experience of the body as a whole (rather than merely the fingers) and also highlights the perceptive faculty of bodily kineasthesics (the body's locomotion in space). This encompasses the sensate experience of the individual's moving body, and the individual's perceptual experience of the moving bodies of others. Following this, within a live performance moment there comes about a 'reciprocity of the haptic system' of perception (Rodaway, 1994: 44) through this experience of tactile and kinaesthetic moments, whether actual or observed.

9 The (syn)aesthetic-sense defines the intuitive human sense that 'presents the unpresentable' (Kant, 1978: 35), enabling a 'sensing beyond' (after Nietzsche, 1967: 132). It encompasses the *noetic* (from the Greek *nous* meaning 'intellect' or 'understanding'), a 'knowledge that is experienced directly', which can provide 'a glimpse of the transcendent' (Cytowic, 1994: 78). The noetic has an 'ineffable quality' in that it makes manifest a complex experience that defies explanation, 'that which by definition cannot be put into words' (Cytowic, 1994: 119–21). See Machon, 2009 for a more detailed explanation.

10 In the immediate moment this speech covers the evolutionary survival of the fittest – cockroaches, wasps, crocodiles, humans, and the twenty-first-century 'apocalypse' of humans against nature, confronting the danger in separating humanity from the primordial. Most disturbingly this speech exposes the recurrence of culpability, drawing attention to the play's own form within that recurrence (the play begins with Harper ensnaring herself in her own chilling lies to the younger Joan). It makes tangible the fact that humans collude with acts of atrocity against other humans, other species and are culpable in globalization, culpable in pollution and climate change, culpable in ethnic cleansing, culpable in lying to children and in believing their own lies. This short speech enables, in the fleeting moment of immediate delivery, and in the subsequent processes of visceral interpretation, a disquieting *re*cognition that, like the chaos theory, the act of lying to a child can produce a world that destroys itself.

References

Artaud, Antonin. (1974). *Collected Works: Volume Four.* Trans. Victor Corti. London: Calder & Boyars.

——. (1993). *The Theatre and its Double.* Trans. Victor Corti. London: Calder Publications.

Broadhurst, Susan. (1999a). *Liminal Acts: A Critical Overview of Contemporary Performance and Theory.* London: Cassell.

——. (1999b). 'The (Im)mediate Body: A Transvaluation of Corporeality'. *Body & Society*, 5.1. London, Thousand Oaks, CA, and New Delhi: Sage Publications: 17–29.

Churchill, Caryl. (1988). Writer. *Fugue.* Director and Choreographer: Ian Spink. Music: J. S. Bach. Musical Director: Nicholas Carr. Dance-Lines Production. Channel 4, London.

——. (1993). *Not ... not ... not ... not enough oxygen and Other Plays*. Ed. Terry Gifford and Gill Round. Essex: Longman.
——. (1994). *The Skriker*. London: Nick Hern Books.
——. (1996). *Plays: 2*. London: Methuen.
——. (1997a). *Blue Heart*. London: Nick Hern Books.
——. (1997b). Writer. *Hotel*. Archive video. Director and Choreographer: Ian Spink. Music: Orlando Gough. Design: Lucy Bevan. Video director: Francesca Penzani. The Place, London.
——. (1997c). *Plays: 1*. London: Methuen.
——. (1998). 'The Lives of the Great Poisoners',. *Plays: 3*. London: Nick Hern Books, pp. 184–237.
——. (2000a). *Far Away*. London: Nick Hern Books.
——. (2000b). Writer. *Far Away*. Director: Stephen Daldry. Designer: Ian McNeil. Royal Court Theatre Upstairs, London. 13 December.
——. (2002a). *A Number*. London: Nick Hern Books.
——. (2002b). Text. *Plants and Ghosts*. Choreography: Siobhan Davies. Movement material: The Siobhan Davies Company dancers. Sound installation: Max Eastley. Lighting Design: Peter Mumford. Costumes: Genevieve Bennett and Sasha Keir. Voiceover: Linda Bassett. Victoria Miro Gallery, London. 16 October.
——. (2006a). *Drunk Enough To Say I Love You?* London: Nick Hern Books.
——. (2006b). Writer. *Drunk Enough To Say I Love You?* Director: James Macdonald. Design: Eugene Lee. Lighting: Peter Mumford. Performers: Ty Burrell and Stephen Dillane. London, Jerwood Theatre Downstairs, The Royal Court. November.
——. (2009). *Seven Jewish Children: A Play for Gaza*. London: Nick Hern Books.
Churchill, Caryl and David Lan. (1998). 'A Mouthful of Birds', *Plays: 3*. Caryl Churchill. London: Nick Hern Books, pp. 2–53.
Cousin, Geraldine. (1989). *Churchill The Playwright*. London: Methuen Drama.
Cytowic, Richard. E. (1994). *The Man Who Tasted Shapes*. London: Abacus.
Derrida, Jacques. (1978). *Writing and Difference*. Trans. Alan Bass. London: Routledge.
——. (1976). *Of Grammatology*. Trans. Gayatri Chakravorty Spivak. Baltimore, MD, and London: The Johns Hopkins University Press.
Diamon, Elin. (1997). *Unmaking Mimesis*. London and New York: Routledge.
Eyre, Richard and Nicholas Wright. (2000). *Changing Stages – A View of British Theatre in the Twentieth Century*. London: Bloomsbury.
Irigaray, Luce. (1985). *Speculum of the Other Woman*. Trans. Gillian C. Gill. Ithaca, NY: Cornell University Press.
Kant, Immanuel. (1911). *Kant's Critique of Human Judgement*. Trans. James Creed Meredith Introductory essays, notes and analytical index James Creed Meredith. Oxford: Clarendon Press.
——. (1978). *Critique of Judgement*. Trans. James Meredith. Oxford: Clarendon Press.
Kristeva, Julia. (1982). *Powers of Horror – An Essay on Abjection*. Trans. Leon S. Roudiez. New York: Columbia University Press.
——. (1992). 'The Novel as Polylogue', in *Desire in Language – A Semiotic Approach to Literature and Art*. Ed. Leon S. Roudiez. Trans. Thomas Gora, Alice Jardine and Leon S. Roudiez.. Oxford: Blackwell, pp. 159–209.
Machon, Josephine. (2009). *(Syn)aesthetics – Redefining Visceral Performance*. Basingstoke and New York: Palgrave Macmillan.

Nietzsche, Friedrich. (1967). *The Birth of Tragedy and The Case of Wagner*. Trans. Walter Kaufmann. New York: Vintage Books.
Novarina, Valère. (1996). *The Theater of the Ears*. Trans. and ed. Allen S. Weiss. Introduction Allen S. Weiss. Los Angeles: Sun & Moon Press.
Rodaway, Paul. (1994). *Sensuous Geographies – Body, Sense and Place*. London and New York: Routledge.
Selden, Raman, Peter Widdowson and Peter Brooker. (1997). *A Reader's Guide to Contemporary Literary Theory*, 4th edn. London: Prentice Hall Harvester Wheatsheaf.
Tozer, Kathy. (2001). Unpublished personal interview. 12 January.

Index